# Measuring Operations Performance

More information about this series at http://www.springer.com/series/13500

Iñaki Heras-Saizarbitoria
Editor

# ISO 9001, ISO 14001, and New Management Standards

Springer

*Editor*
Iñaki Heras-Saizarbitoria
Department of Management
University of the Basque Country
Donostia-San Sebastián
Spain

ISSN 2363-9970 ISSN 2363-9989 (electronic)
Measuring Operations Performance
ISBN 978-3-319-65674-8 ISBN 978-3-319-65675-5 (eBook)
https://doi.org/10.1007/978-3-319-65675-5

Library of Congress Control Number: 2017951178



Printed on acid-free paper

This Springer imprint is published by Springer Nature
The registered company is Springer International Publishing AG
The registered company address is: Gewerbestrasse 11, 6330 Cham, Switzerland

# Preface

This edited book, *ISO 9001, ISO 14001, and New Management Standards*, published by Springer in the "Measuring Operations Performance series", focuses on Management System Standards, also referred to as Certifiable Management System Standards or meta-standards.

These standards are voluntary reference standards used by organizations to formalize, systematize and legitimize a very diverse set of managerial activities and tasks (Boiral and Heras-Saizarbitoria, 2015). The standards have been adopted by an increasing number of organizations across the world. They encompass a very wide range of aspects of business activity, including quality management (e.g. ISO 9001), environmental management (e.g. ISO 14001), the prevention of occupational hazards and the provision of health and safety regulations in the workplace (e.g. OHSAS 18000, ISO 45001) and corporate social responsibility (e.g. SA8000).

This monograph, that has been promoted by the *Research Network on Management Standards (RMS)*, aims to make a substantive contribution to the field of Management System Standards.

Chapter "Three Decades of Dissemination of ISO 9001 and Two of ISO 14001: Looking Back and Ahead" contains a general overview and a descriptive analysis of the dissemination of the two main global management system standards or meta-standards—ISO 14001 and ISO 9001—on an international level.

Chapter "What Drives the Quality of Certifiable Management System Standards Implementation? Insights from the ISO 9001 Standard" sheds light on the drivers related to low, medium and above average adoption of the ISO 9001 standard, using a novel dataset of 201 Greek firms that are ISO 9001 certified.

Chapter "ISO 14001 Adoption and Environmental Performance: The Case of Manufacturing in Sweden" analyzes the adoption of the ISO 14001 standard. Among other issues it analyzes the impact of ISO 14001 on concrete environmental performance, based on empirical work carried out with Swedish manufacturing firms.

Chapter "Adopting ISO Management Standards in Africa: Barriers and Cultural Challenges" highlights the barriers to the adoption of ISO management standards in the African context and also examines whether these standards are adequate to meet the specificities of African cultures and organizational practices.

Chapter "Implementing Service Management Standards: Motivations and Key Factors" analyzes the motivations and key factors in the adoption of the ISO 20000 standard for service management systems.

Chapter "The Internalization of a Sectorial Standard for Quality Management: A Qualitative Analysis in Tourism" analyzes the drivers and the process of internalization of a Spanish sectoral certifiable quality standard, based on a qualitative study.

Chapter "Drivers, Obstacles and Benefits of the Adoption of SA8000: A Survey in Italian Companies" analyzes the adoption of the SA8000 standard, one of the most popular and widespread certifiable standards of the field of corporate social responsibility, based on a quantitative field-work carried out with 130 Italian organizations.

Chapter "Adopting ISO/TS 16949 and IATF 16949 Standards: An Exploratory and Preliminary Study" analyzes the critical points of working with a management system based on ISO/TS 16964 or IATF 16949, considering the strengths and flaws in the adoption of the standard, based on a qualitative exploratory study.

Chapter "Implementing the ISO 50001 System: A Critical Review" analyzes the relevance of the ISO 50001 standard for reducing energy consumption, based on a review of the existing literature, and identifies its limitations, advantages and possible weaknesses.

Chapter "Conflicts Arising in the Generation Process of the ISO 45001 Standard" sheds light on the process involved in generating the ISO 45001 standard in the field of Occupational Health and Safety.

Finally, Chapter "A Supply Chain View on Certification Standards: Does Supply Chain Certification Improve Performance Outcomes?" examines the perceived performance implications of ISO 9001, ISO 14001, and OHSAS 18001 at the supply chain level.

I would like to thank all the contributors to this volume most sincerely, including the authors whose chapters were accepted for inclusion, and the authors who submitted that could not be included because of limitations on space. My special thanks go to Prof. Dr. Andrea Chiarini for his invitation to edit this monograph and to Ms. Maria Cristina Acocella from Springer for her kind support from the very beginning. I am also indebted to all the reviewers that collaborated in the review process.

Donostia-San Sebastián, Spain
June 2017

Iñaki Heras-Saizarbitoria

## Reference

Boiral O, Heras-Saizarbitoria I (2015) Management system standards, meta perspective. In Dahlgaard-Park SM (ed) (2015), The sage encyclopedia of quality and the service economy, Sage Publications, Thousand Oaks, CA

# Contents

# Three Decades of Dissemination of ISO 9001 and Two of ISO 14001: Looking Back and Ahead

**Iñaki Heras-Saizarbitoria, Olivier Boiral and Erlantz Allur**

**Abstract** This introductory chapter contains a general overview and a descriptive analysis of the dissemination of the two main global management system standards or metastandards—the ISO 14001 and the ISO 9001—on an international level. The first two global certifiable management system standards are presented, together with a descriptive analysis of the global dissemination of both. Similarly, insights related to some of the conclusions of the scholarly literature on the subject and a short reference to avenues for further research are also summarized in this chapter.

**Keywords** Management system standards · Metastandards · Standardization · Self-regulation · ISO 9001 · ISO 14001 · Dissemination

## 1 Introduction

Standardization has been crucial for the development of the industrial society (Brunsson and Jacobsson 2000; Blind 2004). At its origins, in the early twentieth century, standardization was introduced in order to curb an uneconomical divergence of components, parts and supplies and to foster their interchangeability so as to facilitate mass production and the repair and maintenance of products and services. Standardization, however, has gone further than this and has come to be applied to the very management processes and systems by which products and

I. Heras-Saizarbitoria (✉) · E. Allur
Department of Business Management, University of the Basque Country, San Sebastian, Spain
e-mail: iheras@ehu.es

E. Allur
e-mail: erlantz.allur@ehu.es

O. Boiral
Department of Management, Faculty of Business Administration, Université Laval, Quebec City, Canada
e-mail: Olivier.Boiral@mng.ulaval.ca

 
I. Heras-Saizarbitoria (ed.), *ISO 9001, ISO 14001, and New Management Standards*, Measuring Operations Performance, https://doi.org/10.1007/978-3-319-65675-5_1

services are produced. Brunsson and Jacobsson (2000) refer to these as *standards for administrative processes*, and Furusten (2000, p. 71) as "*standards on how to design and manage organizations*". These *administrative standards* are more frequently referred to as management system standards (MSSs), a term that already has a certain tradition of academic acceptance (e.g. Delmas 2002; Neumayer and Perkins 2004), and also as to metastandards (e.g. Uzumeri 1997; Heras-Saizarbitoria and Boiral 2013). In the last four decades there has been and intensive acceleration of the process of standardization with regard to business management, in an economic context characterized by a marked process of globalization and economic integration of markets.

MSSs are voluntary codes, guidelines or processes used by organizations to formalize, systematize, and legitimize a very diverse set of managerial activities or tasks (Boiral and Heras-Saizarbitoria 2015). These MSSs are, of course, to be distinguished from the technical norms and specifications relating to those requirements with which particular products or processes need to comply. These are standards which correspond to the standardization of a very wide range of aspects of business activity, such as quality management (e.g. ISO 9001), environmental management (e.g. ISO 14001), the prevention of occupational hazards and the provision of health and safety regulations in the workplace (e.g. OHSAS 18001 and ISO 45001), and corporate social responsibility (e.g. SA8000, ISO 26000). All of these standards tend to have a very similar methodology in relation to their creation, structure, process of implementation and monitoring by a third party.

The remainder of the chapter is organized as follows. Following this introduction, in the second and the third sections, ISO 9001 and ISO 14001—the first two global MSSs—are presented, respectively. The fourth section contains a descriptive analysis of the global dissemination of both MSSs. The chapter concludes with a summary of the main discussion and the conclusions drawn, together with the mentioned insights about the outcomes of the scholarly literature and the reference to the avenues for further research.

## 2 ISO 9001: The First International Management Standard

ISO 9001 is arguably the most influential single international MSSs that there has been to date (Braun 2005). The ISO 9000 family of standards was created, in its initial form, in 1987, and underwent substantial revisions in 1994 and 2000.[1] The

[1] "ISO 9000 standards" or "the ISO 9000 family of standards" are the expressions commonly used to refer to the totality of standards in the same series, although in the 2000 version the only standard which includes a model for implementing a certifiable management system (i.e., the only management system standard) is the ISO 9001 standard. In fact, in the 1987 and 1994 versions, in addition to the ISO 9001 standard, the ISO 9002 and ISO 9003 standards also included certifiable management system models, which is why the plural form used previously continues to be

last version of the standard was launched in 2015. On a global level these standards spread in their initial phase throughout the countries of the E.U., becoming particularly prevalent in the UK, which is perfectly logical in view of that country's previous experience with the BS 5750. It should also be borne in mind, moreover, that E.U. institutions, and specifically the European Commission, promoted intensively the adoption of this standard by European companies, as part of the process of harmonization that was established with a view to creating the single European market in 1992 (Tsiotras and Gotzamani 1996), and that it was even included in the commercial directives of what was then the European Community (Anderson et al. 1999; Mendel 2002).

On the other hand, the adoption of ISO 9001 was much less intensive in the USA and Japan. In fact they came in for considerable criticism, and were initially considered as clear non-tariff barriers in those countries. Nevertheless, in the last two decades it has been a significant increase in their use. First, companies exporting to the E.U. have been fostered to obtain certification. Second, certain key institutional organisms in the two countries have adopted and promoted the implementation of these standards.[2] With the EC's acceptance of ISO 9001 certification, other governments adopted it for the very beginning. U.S. government agencies that have adopted ISO 9001 include the following: the U.S. Departments of Defense and Energy, the Food and Drug Administration and the Federal Aviation Administration (Anderson et al. 1999).

It has to be pointed out that ISO 9001 standard are not standards that refer to compliance with an objective or with a particular result. ISO 9001 is not a performance standard to measure the quality of companies' products or services. It is a standard that establish the need to systematize and formalize a whole series of company processes into a series of procedures, and to document this implementation. ISO 9001 standardizes procedures, duties, and roles, rather than goals or outcomes (Braun 2005).

In short, compliance with ISO 9001—a fact which is certified by an organization accredited for this purpose—means having documentation to show the implementation of a quality management system which includes in standardized and documented procedures the basic processes used to produce the product or service which the customer acquires. These standards are a management tool based on the systematization and formalization of tasks in order to achieve product homogeneity and to conform to the specifications established by the customer (Anderson et al. 1999).

(Footnote 1 continued)

employed. Loosely speaking, reference is still made to "ISO 9000" or "certified in accordance with ISO 9000", whereas, properly speaking, the correct expression should be "ISO 9001 standard" or "certified in accordance with ISO 9001" (there is an ISO 9000 standard in the current series, but it is a standard applied to definitions and terminology).

[2]For example, such important public organisms as the U.S. Department of Defense or the very influential Food and Drug Administration (FDA), together with other organisms of a private nature, such as the association of chemical manufacturers or the association of automotive industry manufacturers, all adopted the ISO 9001 standard.

In other words, as one manager summarized to Cole (1999), "document what you do, do what you document, and verify that you are doing it" (Cole 1999; 114). Such a clarification is, in our opinion, especially pertinent, since there have been major misunderstandings in this respect on numerous occasions in the past, and in a variety of different fields. For example, in countries such as Spain and Italy—leaders in the world league table for ISO 9001 and ISO 14001 certification—numerous companies have publicized their certificate as a registered company as though it were a product quality label, despite the fact that the ISO standard itself prohibits such a practice. In the scholarly literature this issue has been analyzed, for example for the case of Malaysia, Abdullah et al. (2009) evidenced that about 44 and 51% of the advertisements sampled fall under the very high to high level and moderate levels of non-compliance categories, respectively.

It should also be stated that the implementation of this type of standard or norm is voluntary, although in certain sectors their application constitutes a de facto obligation. In this way (and as will be examined subsequently below), in those studies in which an analysis has been made of companies' motivations for obtaining certification, considerable emphasis has been accorded to the "prescriptive" role played by large companies in the construction, automotive, energy and telecommunications sectors. The latter saw in the ISO 9001 standard a way of ensuring a certain level of quality from their suppliers and subcontractors, in the sense of obtaining a certain systematization and formalization of the key processes utilized by such companies to comply with the requirements that the larger companies had established, but without increasing their operational costs.

## 3 ISO 14001: The Global Green Standard

The ISO 14001[3] standard was promulgated in 1996, benefiting from the success enjoyed by the ISO 9000 family of standards. ISO 14001 was based on the model of various national environmental management standards, and in particular on the British BS 7750 standard.

The standard was created in the institutional context of the 1990s, a period characterized by the strengthening of the trend towards seeking a green paradigm for production and consumption. Motivated by different stakeholders and the internal improvement of their general and environmental efficiency, an increasing number of companies began to introduce at this time an Environmental Management System (EMS). An EMS is a systematic process that corporations and other organizations use in order to implement environmental goals, policies and

[3]By analogy with the case of the ISO 9000 standards, it is also commonplace to talk of the "ISO 14000 standards" to refer to the totality of the standards in this series, although the only standard which includes a model for implementing a certifiable management system is the ISO 14001 standard. In this case, moreover, it should be borne in mind that there is no ISO 14000 standard, properly speaking.

responsibilities, as well as to provide for regular auditing of these aspects (Cascio 1996).

For Haufler (1999), some companies introduced an EMS so as to avoid the introduction of more far-reaching public environmental regulations, while others introduced them to respond to the criticisms made by environmental activists and yet others adopted them so as to enhance the efficiency and sustainability of their businesses. Some stakeholders, including those who were most critical and those responsible for public regulation, greeted the promulgation and implementation of these EMS standards with scepticism, principally due to the difficulty of knowing what is involved by the fact that a company has introduced an EMS, and how precisely its environmental performance improves as a result of its introduction (Boiral et al. 2017).

The creation of an ISO standards from the ISO 14001 series represented a new feature, albeit one not without controversy, given that for the first time reference was being made to issues of a political and social nature—aspects about which for many the International Standardization Organization had neither legitimacy nor authority. Apart from the criticism regarding its controversial creation process, the ISO 14001 standard has been subject to very diverse criticism in terms of both its content and the procedure established for its certification (Haufler 1999). Essentially, this criticism revolved around the fact that the standard was not geared towards demanding an improvement in environmental performance, in the sense that companies are not required to attain certain environmental results, or to ensure consistency in terms of their external auditing service (King et al. 2005).

It has to be pointed out, once again, that this standard does not fix environmental goals or environmental targets to be achieved (requirements for the prevention and reduction of the impact of pollution, for instance), as a result of the possible attainment of which a certificate would be obtained. Rather, these standards establish requirements defining the operational systems to be complied with within companies in relation to activities which have an environmental impact. In short, it is a model which provides a systematic framework within which to incorporate environmental concerns into a company's day-to-day operations.

## 4 Global Impact of ISO 9001 and ISO 14001

A comparative analysis of the global impact of ISO 9001 and ISO 14001 certificates is presented in this section. An index of our own design is proposed for such purpose: the Intensity of Certification Index (ICI). ICI measures the relationship between the world percentage of certificates in each country with respect to each country's percentage contribution to the world GDP. The source of the certification data is the proper ISO, who every year publishes a survey with statistics regarding the dissemination of ISO 9001 and ISO 14001 all over the world. As underlined by the proper ISO, these data are prone to several biases; among many other the following two limitations could be mentioned (ISO 2016): first, there is a variability in numbers of certificates reported each year by individual certification bodies; second, there is an inconsistent participation of some certification bodies that contribute to the survey 1 year but not the next.

By late 2015, over 1,033,936 ISO 9001 certificates had been authorized, thus more than 625,000 the number of certificates compared with the figure for the end of the year 2000, a year during which a new version of the standards was launched, and by the end of which there were a total of 408,631 such certificates (ISO 2016) (Fig. 1).

Continent by continent, it can be seen that the EU, with 439,477 certificates issued, continues to lead in terms of total number of ISO 9001 certificates, since it absorbs nearly half the total number of certificates awarded throughout the world (more precisely, 42.50% of the total). Nevertheless, this global leadership has diminished by more than four percentage points when compared with the year 2008. China is the country with the largest number of certificates in the world (having a total of 292,559 by the end of 2015), followed by Italy (with 132,870), Germany (52,995), Japan (47,101) and United Kingdom (40,161) (Table 1).

On the other hand, ISO 14001 certification has been experiencing major growth on the international stage in recent years. We only have to highlight the fact that, if by the end of the year 1999 13,994 certificates had been issued worldwide, by the end of 2015 the number of certificates issued had reached 319,324. In other words, in the space of 16 years, the number of certificates issued worldwide had increased more than 22-fold.

Attention should be drawn to the fact that around 40% of ISO 14001 certificates issued worldwide were issued within the EU. The USA's share, on the other hand, was limited to 1.9%, while China and Japan were undeniably the world leaders in terms of the absolute number of certificates issued, accounting for 36 and 8% of the total respectively. In recent years, special mention should be made of the growth in

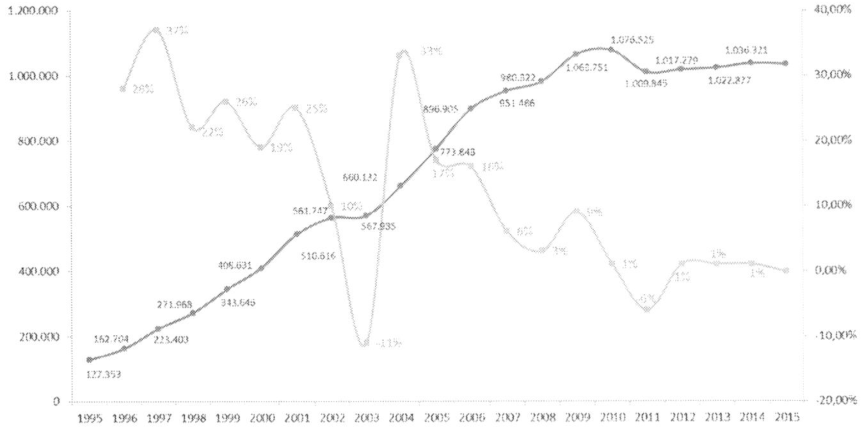


**Fig. 1** Worldwide temporal dissemination of ISO 9001 (1995–2015). *Source* Own preparation on the basis of the various executive reports on the global statistics for ISO 9001 published by ISO

**Table 1** International adoption of ISO 9001 in the main global economic zones

| | 1997 | 2001 | 2008 | 2015 | 2015 | |
|---|---|---|---|---|---|---|
| | | | | | % | ICI |
| USA | 18,581 | 37,026 | 32,400 | 33,103 | 3.26 | 0.13 |
| Japan | 6487 | 27,385 | 62,746 | 47,101 | 4.65 | 0.77 |
| European Union | 135,984 | 253,488 | 401,239 | 439,477 | 42.50 | 1.80 |
| China | 5698 | 57,783 | 224,616 | 292,559 | 28.87 | 1.90 |
| World | 223,298 | 510,349 | 980,322 | 1,033,936 | 100 | 1 |

*Source* Own preparation on the basis of the various executive reports on the global statistics for ISO 9001 published by ISO

*Note* Up to the year 2000 the certificates analysed are ISO 9001, ISO 9002 and ISO 9003; from 2001 onwards the certificate analysed is ISO 9001:2000. ICI calculated as the ratio between the percentage share of the number of certificates issued worldwide and the percentage share of total world GDP PPP in 2015 measured in US dollars

**Table 2** International adoption of ISO 14001 in the main global economic zones

| | 1999 | 2001 | 2008 | 2015 | 2015 | |
|---|---|---|---|---|---|---|
| | | | | | % | ICI |
| USA | 636 | 1645 | 4974 | 6067 | 1.92 | 0.08 |
| Japan | 3015 | 8123 | 35,573 | 26,069 | 8.24 | 1.36 |
| European Union | 7253 | 17,941 | 78,118 | 119,754 | 37.50 | 1.30 |
| China | 222 | 1085 | 39,195 | 114,303 | 36.13 | 2.37 |
| World | 13,994 | 36,464 | 188,574 | 319,324 | 100 | 1 |

*Source* Own preparation on the basis of the various executive reports on the global statistics for ISO 14001 published by ISO. IC-GDP: ICI calculated as the ratio between the percentage share of the number of certificates issued worldwide and the percentage share of total world GDP in 2015 measured in US dollars at current exchange rates of the time (World Bank)

the number of certificates issued in the People's Republic of China (see also, Boiral et al. 2017) (Table 2; Fig. 2).

The growth in the numbers of certificates when broken down according to world region suggests a roughly similar pattern of geographical distribution for ISO 9001 and ISO 14001 (Marimón et al. 2006).

In short, and looking at different regions of the world, Europe leads the way in adoption of both standards, with East Asia second and North America third. On the contrary, many countries in Western and Southern Asia and especially in Africa—with the notable exception of South Africa—still lag behind in certification numbers, both for ISO 14001 and ISO 9001. This pattern is clearly visible in Figs. 3 and 4.

Within the EU-27 (see Fig. 5), with the ICI calculated as being the ratio between the percentage share of the number of certificates issued in the EU and the percentage share of total GDP of the EU, of special note is the loss in intensity of the United Kingdom and the growth experienced by Italy and Spain—two leading countries as far as the implementation of MSSs is concerned. Specifically and in absolute terms, Italy is ranked second in the world in terms of the number of ISO

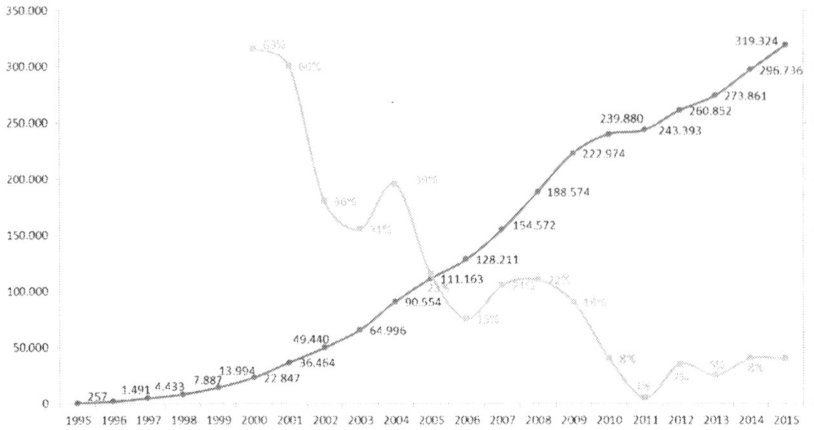


**Fig. 2** Worldwide temporal diffusion of ISO 14001 (1995–2015). *Source* Own preparation on the basis of the various executive reports on the global statistics for ISO 9001 published by ISO

9001 certificates (with 132,870 certifications by the end of 2015), followed by Germany (with 52,995), UK (40,161) and Spain (32,730). As for ISO 14001, Italy ranks third (with 22,350 certifications by the end of 2015), preceded only by China and Japan, followed by UK (17,824) and Spain (13,310). Special mention should also be made of the performance of some of the countries that have most recently joined the Union (e.g. Hungary, Rumania), which have attracted considerable investment in industrial production and have been experiencing growth levels way above the average for EU-27 countries in terms of the number of both ISO 9001 and ISO 14001 certificates issued. On the other hand, the Nordic countries evidence a low level of intensity in terms of ISO 9001 certifications while, on the contrary, experience among the highest levels of intensity of ISO 14001 certification for the 1997–2008—not surprisingly, since they have historically been among the most environmentally proactive countries (Figs. 6 and 7).

## 5 Discussion and Conclusions

The successful diffusion of both ISO 9001 and ISO 14001 might be related to the basic impetus of the process of globalization of the Western economies, to the extending global supply chains and the crucial role of transnational corporations. In the current economic environment, in which outsourcing and relocation of companies' activities have become key strategic elements of global supply chains, it is necessary to foster a certain level of homogeneity of management systems in order to favour the development of such processes, and metastandards may help to

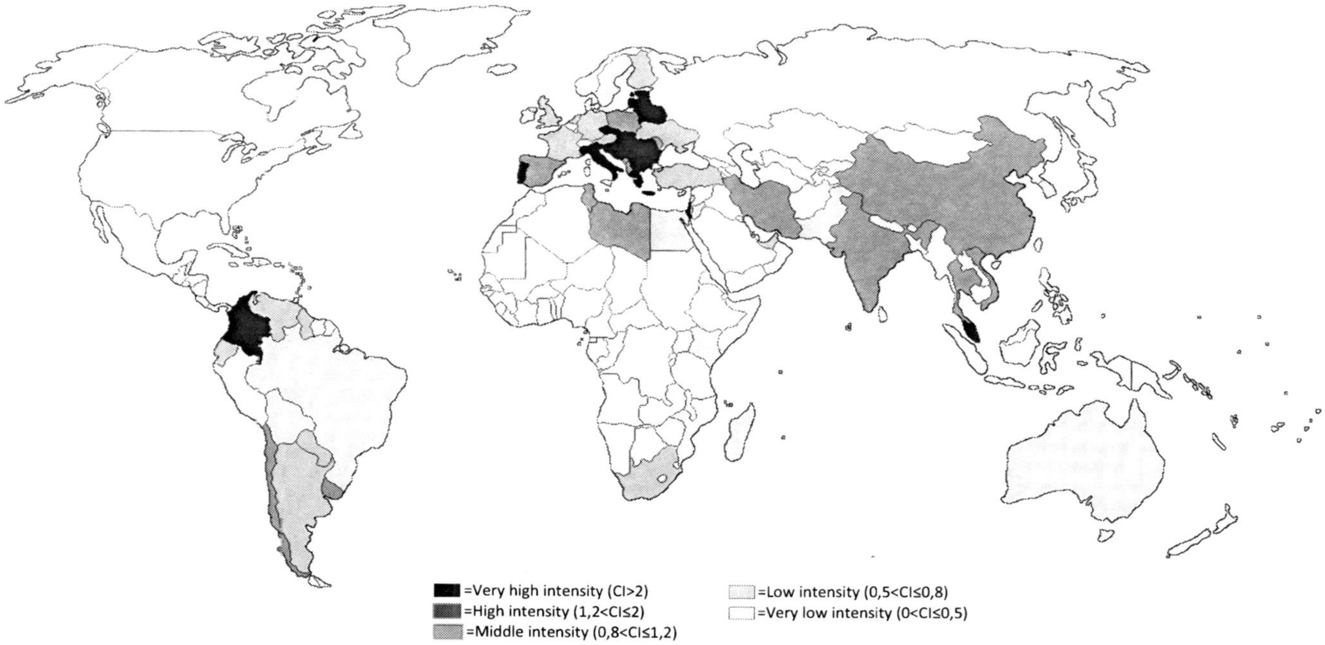


**Fig. 3** Global intensity of ISO 9001 certification. *Source* Own preparation on the basis of ISO reports on worldwide certification levels and of World Bank data for GDP both globally and broken down by country

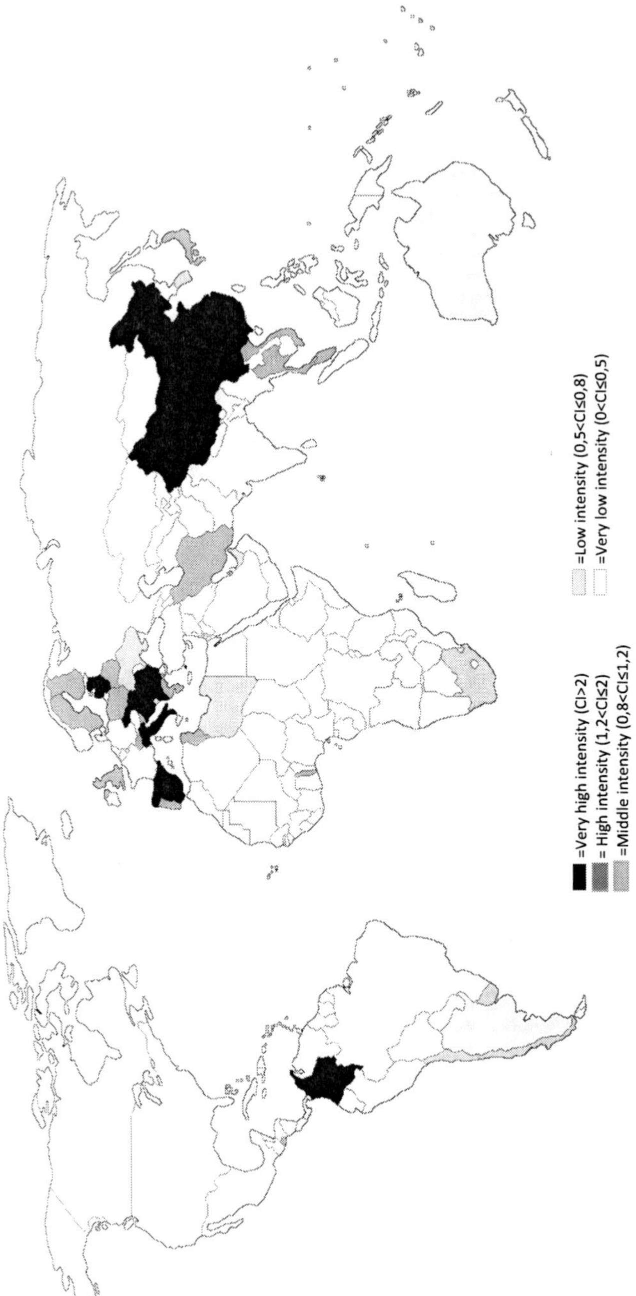


**Fig. 4** Global intensity of ISO 14001 certification. *Source* Own preparation on the basis of ISO reports on worldwide certification levels and of World Bank data for GDP both globally and broken down by country

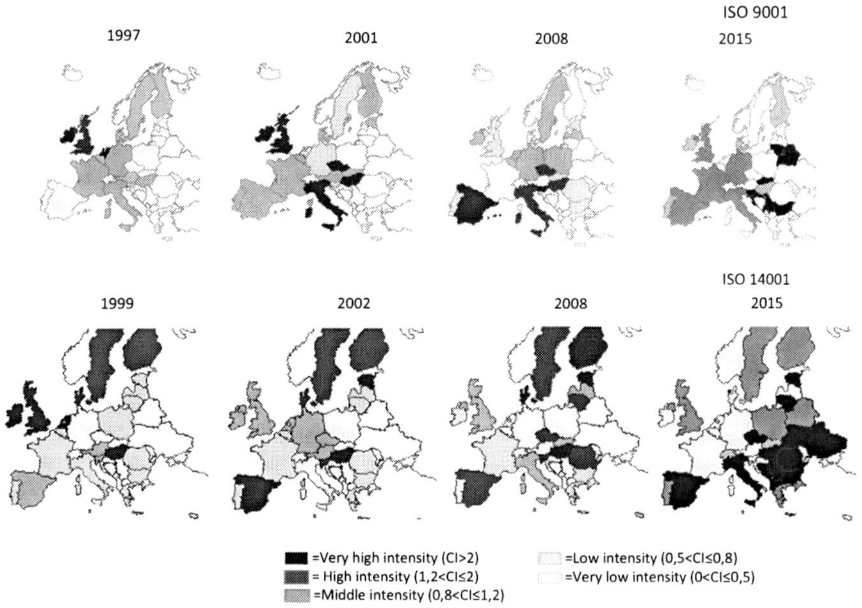


**Fig. 5** Evolution of the intensity of certification of ISO 9001 and ISO 14001 in the EU (1997–2015). *Note* Calculations made on the basis of GDP for 2004 measured in terms of PPP (purchasing power parity). *Source* Own preparation on the basis of data obtained from ISO reports and from Eurostat

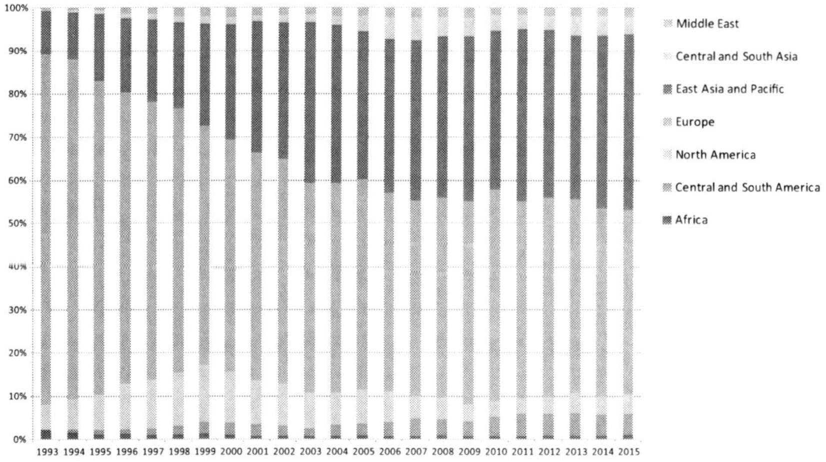


**Fig. 6** Regional share of ISO 9001 certification (1993–2015). *Source* Own preparation on the basis of ISO, 2016

achieve this aim. The generalization of MSSs across the world has also been perceived by international institutions such as the European Union and the World Trade Organization as a way to reduce technical barriers to trade, facilitate commercial transaction and reinforce the process of globalization.

The major level of dissemination achieved by both ISO 9001 and ISO 14001 has to be pointed out despite their unequal dissemination has to be also highlighted. There is a clear predominance of countries with disparate market structure and institutional environment, such as Italy, Spain and China. Conversely, other regions of the world are very far away from the successful dissemination of the main global standards. As underlined in the chapter of this volume by Tene et al. (2017) the number of certifications in the African continent increased by 20% for ISO 9001 and 19% for ISO 14001. Despite this rapid increase in absolute terms, ISO certifications in Africa remain relatively underdeveloped with about 1% of the total number of ISO 9001 and ISO 14001 certifications worldwide This is nothing but another indicator of the unbalanced world economic development: as most areas of Africa are outside the global system of production and the global supply chains, the incidence of both ISO 9001 and ISO 14001 standards is really low.

Although many interesting results have been obtained from the scholarly research regarding MSSs—see, for this purpose, the other chapters of this volume published by Springer—, clear answers are still waiting to be found. Regarding the outcomes of these standards, despite many scholarly works have been carried out in the last decades (see, for two recent reviews, Heras-Saizarbitoria and Boiral 2013; Boiral et al. 2017) responses to such elementary questions as to what we can attribute the fact that some countries which are not present at the highest level in international indices of competitiveness (e.g. such as Italy, Spain and Portugal) nevertheless have indices of ISO 9001 certification in relation to their economic level that place them at the top of the global league table, whereas other countries that appear at the top of the competitiveness indices referred to even within the E.U. (e.g. Germany and Finland) have much lower indices of certification. A similar remark could be established regarding the ISO 14001 certification intensity and the concern for environmental issues of the leading countries.

Similarly, as underlined by Boiral and Heras-Saizarbitoria (2015) despite the practitioner literature tends to underline MSSs benefits, more academic and critical research has showed that their outcomes are far from automatic and foreseeable. These outcomes are notably related to the attitudes towards the MSSs of the adopting organizations and the way they are implemented: as these standards are often adopted as a result of external pressures, the most essential question is not whether certification should be contemplated and what impact can be expected, but rather how MSSs may be used as efficiently as possible. Regarding the main pitfalls, the risks of too much procedures and formalization have to be underlined, together to the lack of internalization of the standards. In the scholarly field there is a knowledge gap in the field of rigorous empirical studies aimed at analyzing the perceptions of the various stakeholders with regard to the process of adoption and internalization of MSSs.

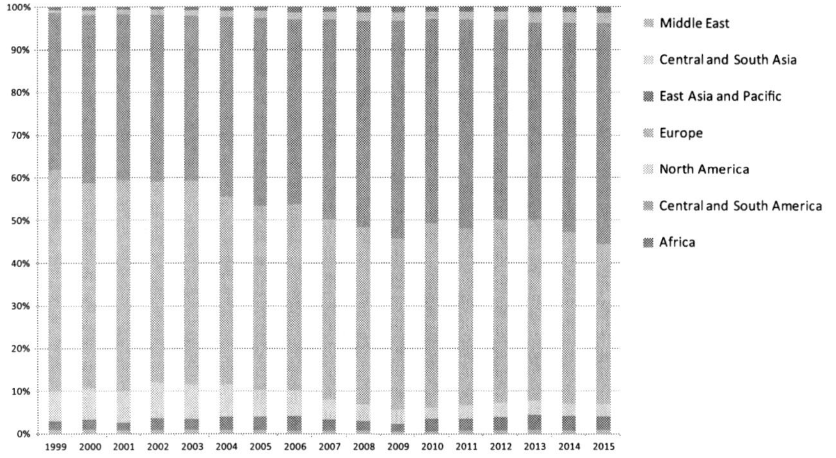


**Fig. 7** Regional share of ISO 14001 certification. *Source* Own preparation on the basis of ISO, 2016

The process of evolution of the scope and content of the main global metastandards ISO 9001 and ISO 14001 should be also analyzed in more depth by scholarly work. Many issues should be analyzed in depth. Among many other, the extent in which these revision processes are guided by real needs for substantive improvements for the standards or by commercial purposes. The recent work by Anttila and Jussila (2017) focused on the process of design and launching of the 2015 version of the ISO 9001 standard (the so-called ISO 9001:2015 standard) raised deep and substantive concerns about the revision of the standard that could be analyzed in more depth. Similarly, the real implication of different stakeholders in the process of revision of metastandards should be also analyzed.

Finally, although the external recognition and social legitimacy of the MSSs is largely taken for granted in the literature, the credibility and value added of these standards in the eyes of different stakeholders (consumers, managers, suppliers, intermediary clients, workers, public administration, etc.) remain unclear. Future research could also focus on third-party auditing, in order to analyse in depth the consistency of external auditing services that have been questioned in this field as in other areas of business and management, notably in the accountancy field. The existence of many entities capable of performing the required audits and issuing the relevant certificates was pointed out as a weakness that favoured a symbolic and instrumental implementation of MSSs, contributing to erode their prestige and credibility. Generally speaking, certifiable MSSs seem to have contributed to substantially reduce the number of audits conducted by clients companies to verify suppliers' and subcontractors' procedures. Nevertheless, certain concerns have been underlined in recent years regarding to the reliability of the whole system of standardization in some specific regions and countries such as China, Pakistan or

Russia. An independent and rigorous scholarly research is needed in order to shed light on these and other issues related to the global dissemination of metastandards.

**Acknowledgements** This chapter is a result of a Research Group funded by the Basque Autonomous Government (Grupos de Investigación del Sistema Universitario Vasco; IT1073-16).

## References

Abdullah HS, Mustapha RMBR, Kaliannan M et al (2009) The use (and abuse) of ISO 9000 certification marks in promotional materials in Malaysia. Int J Bus Manag 4(1):168

Anderson SW, Daly JD, Johnson MF (1999) Why firms seek ISO 9000 certification: regulatory compliance or competitive advantage? Prod Oper Manag 8(1):28–43

Anttila J, Jussila K (2017) ISO 9001:2015—a questionable reform. What should the implementing organisations understand and do? Total Qual Manage Bus Excell 28:1–16

Blind K (2004) The economics of standards: theory, evidence, policy. Edward Elgar Publishing, Boston

Boiral O, Heras-Saizarbitoria I (2015) Management system standards, meta perspective. In: Dahlgaard-Park SM (ed) The Sage encyclopedia of quality and the service economy. Sage Publications, Thousand Oaks

Boiral O, Guillaumie L, Heras-Saizarbitoria I, Tayo Tene CV (2017) Adoption and outcomes of ISO 14001: a systematic review. Int J Manage Rev (in press)

Braun B (2005) Building global institutions: the diffusion of management standards in the world economy—an institutional perspective, 3-27. In: Alvstam CG, Schamp EW (eds) Linking industries across the world. Ashgate, London

Brunsson N, Jacobsson B (2000) The contemporary expansion of standardization. In: Brunsson N, Jacobsson B (eds) A world of standards. Oxford University Press, Oxford, pp 1–17

Cascio J (1996) The ISO 14000 handbook. CEEM Information Services, New York, NY

Cole RE (1999) Managing quality fads: how American business learned to play the quality game. Oxford University Press, Oxford

Delmas M (2002) The diffusion of environmental management standards in Europe and in the United States: an institutional perspective. Policy Sci 35(1):91–119

Furusten S (2000) The knowledge base of standards. In: Brunsson N, Jacobsson B (eds) A world of standards. Oxford University Press, Oxford, pp 71–84

Haufler V (1999) Negotiating international standards for environmental management systems: the ISO 14000 standards. UN Vision Project on Global Public Policy Networks, New York, NY

Heras-Saizarbitoria I, Boiral O (2013) ISO 9001 and ISO 14001: towards a research agenda on management system standards. Int J Manage Rev 15(1):47–65

ISO (2001–2016) The ISO survey of ISO 9001 and ISO 14001 certifications. ISO, Geneva

King AA, Lenox MJ, Terlaak A (2005) The strategic use of decentralized institutions: exploring certification with the ISO 14001 management standard. Acad Manag J 48(6):1091–1106

Marimón F, Casadesús M, Heras I (2006) ISO 9001 and ISO 14001 standards: an international diffusion model. Int J Oper Prod Manage 26(2):141–165

Mendel PJ (2002) International standardization and global governance: the spread of quality and environmental management standards. In: Hoffman AJ, Ventresca MJ (eds) Organizations, policy and the natural environment: institutional and strategic perspectives. Stanford University Press, Stanford, pp 407–424

Neumayer E, Perkins R (2004). What explains the uneven take-up of ISO 14001 at the global level? A panel-data analysis. Environ Plann A 36(5):823–839

Tsiotras G, Gotzamani K (1996). ISO 9000 as an entry key to TQM: the case of Greek industry. Int J Qual Reliab Manag 13(4):64–76

Tene CT, Yuriev A, Boiral O (2017) Adopting ISO management standards in Africa: barriers and cultural challenges. In: Heras-Saizarbitoria (ed) ISO 9001, ISO 14001, and new management standards (in press)

Uzumeri MV (1997) ISO 9000 and other metastandards: principles for management practice? Acad Manage Executive 11(1):21–36

# What Drives the Quality of Certifiable Management System Standards Implementation? Insights from the ISO 9001 Standard

**Konstantinos Iatridis and Effie Kesidou**

**Abstract** Despite the important role certifiable management system standards (CMSS) play in our globalised economies, there still exists a gap in the literature regarding firms' motives to implement these standards. This is mainly due to the fact that existing literature perceives certification as synonymous with implementation and does not take into account that quality of implementation, of CMSS among companies, varies. This chapter makes two important contributions. The first one is that it provides quantitative evidence to the relatively few studies that analyse quality of CMSS implementation. In doing so, this work advances our understanding of the factors that motivate firms to implement these standards. Most importantly, in contrast to previous studies that analyse drivers of average implementation of CMSS, this chapter also examines the motives related to low, medium and above average implementation of such standards. Using a novel dataset of 201, ISO 9001 certified firms, located in Greece, our analysis shows that sixty per cent of firms in the sample fail to conform to the requirements of the standard. In contrast to the dominant neo-institutional view in the analysis of CMSS, our econometric findings suggest that competitive motivations are the influential drivers in making companies fully commit to ISO 9001 (i.e. above average implementation).

**Keywords** Certifiable management system standards · Self-regulation · Substantive implementation · ISO 9001 · Corporate social responsibility

K. Iatridis (✉)
School of Management, Centre for Business, Organisations and Society,
University of Bath, Bath, UK
e-mail: k.iatridis@bath.ac.uk

E. Kesidou
Leeds University Business School, University of Leeds, Leeds, UK
e-mail: E.Kesidou@leeds.ac.uk



I. Heras-Saizarbitoria (ed.), *ISO 9001, ISO 14001, and New Management Standards*,
Measuring Operations Performance, https://doi.org/10.1007/978-3-319-65675-5_2

# 1 Introduction

The retreat of the state and the transformation of the world into a global village have caused a dramatic increase in the number of self-regulatory tools, such as Certified Management System Standards (CMSS). ISO 9001, ISO 14001, ISO27001, SA8000 and OHSAS 18001 are all examples of CMSS that have mushroomed over the last twenty five years or so in order to regulate different aspects of business conduct. As a result, more than 1,500,000 companies across 191 countries, nowadays, apply one or more of these standards (ISO 2013). According to the International Standardisation Organisation (ISO Focus 2014), CMSS play a very important role into the economy as, apart from promoting responsible business practices, these management standards aim to increase consumer trust, make supply chains more reliable, and help public authorities to create and implement preventive, deterrent and punitive policies.

Despite the increasing prevalence of CMSS, research into how firms implement these standards remains relatively scarce (Aravind and Christmann 2011; Gondo and Amis 2013; Iatridis et al. 2016), creating a knowledge gap on whether firms fully commit to the CMSS requirements (substantive implementation) or not (symbolic implementation). What is more, the existing literature does not analyse the motivations of substantive implementation of CMSS standards as it mostly focuses on what drives the adoption of CMSS (Heras-Saizarbitoria and Boiral 2013a). However, while firms may adopt CMSS, the quality of implementation could differ among them (Christmann and Taylor 2006). The quality of CMSS implementation could range from those firms that are fully committed to the requirements of such practices, by consistently using the CMSS practices in their daily operations, to those firms that fail to commit to the CMSS, by non-conforming to their proclaimed CMSS values and failing to use the standards' practices in their daily operations. Clearly, there is a need for more studies in this area and this is why scholars have called for more research focussing on the quality of CMSS implementation (Iatridis and Kesidou 2016; Christmann and Taylor 2006; Heras-Saizarbitoria and Boiral 2013a; Nair and Prajogo 2009).

This chapter contributes to the literature that analyses how firms implement CMSS by providing quantitative evidence on two important and under-researched issues: (a) the quality of CMSS implementation, namely, whether companies that adopt CMSS implement them in a substantive or symbolic manner; and (b) the factors that stimulate the substantive implementation of CMSS. We examine these issues by using new primary data of 201 ISO 9001 certified companies in Greece.

The chapter is organized into five sections. The second section provides commentary on CMSS features; it discusses the existing research in CMSS and introduces the research hypotheses. The third section describes the empirical methodology, while the fourth presents and discusses the results. The final section provides the main conclusions and highlights directions for future research.

# 2 Theoretical Background and Hypotheses

## 2.1 Certified Management System Standards

CMSS focus on various aspects of firm's operations, such as quality of products/services (ISO 9001), environmental protection (ISO 14001), information security (ISO 27001) and employee health and safety (OHSAS 18001, SA8000). These standards are the result of a wide multi-stakeholder consultation process in which various groups are involved including business experts, consumer associations, NGOs, academics, governmental authorities and in some cases testing laboratories (European Commission 2012).

CMSS are subject to annual external audits by certification bodies and are applicable to every organization regardless of sector or size. Also, these standards are voluntary and require compliance with relevant regulations. Furthermore, they are process standards and not performance standards meaning that they do not require from organizations a specified environmental or social performance. CMSS require the development and implementation of a structured framework of policies and procedures, which enable the firm to identify and manage its social and environmental impacts in a systematic way (Cragg 2005). Their basic assumption is that better management and documentation of a firm's operational processes and procedures will eventually lead to a better performance (BSI 2013).

Assurance of implementation is succeeded through internal and external audits. The team responsible for implementing the standard carries internal audits at least twice per year. Possible non-conformities with the CMSS requirements are noted and corrective actions are implemented in order to ensure continuous improvement (Behnam and MacLean 2011; Rasche 2009). Additionally, trained and accredited external auditors call for evidence of compliance to a detailed set of requirements in order to certify the firm in question. This certification lasts for three years and then companies need to re-certify their activities. Within the three years of certification, certified firms are audited, at least once per year, externally to ensure compliance with the standards' requirements.

The operating principle, underlying most CMSS, is the Deming Cycle, otherwise known as the Plan-Do-Check-Act (PDCA) Cycle (Deming 1982). This identifies a certain management strategy for certified firms in order to continuously improve their performance with respect to the issue addressed by the standard. According to this principle, first, firms must analyse their current position, set objectives and targets and then make plans to achieve them (Plan); second, they must put these plans into action (Do) by using daily the documents of the management system, created for the purpose of the standard; third, they have to measure their performance against the set objectives and targets through internal and external audits (Check); and, fourth, in case of any major malfunctions or declinations from the standard's requirements, otherwise known as major non-conformances, they have to apply the required corrective actions to improve any flaws (Act) (ISO 2009).

## 2.2 *Quality of Certified Management System Standards Implementation*

The majority of existing research treats certification as a reliable measure of CMSS implementation (Heras-Saizarbitoria and Boiral 2013a, b; King et al. 2005). Albuquerque et al. (2007) for instance, look at the diffusion of the ISO 9001 and ISO 14001 quality and environmental management systems standards respectively, but always treat adoption of CMSS as a binary variable. Equally, Delmas and Montel (2008) offer valuable insights into the same topic but they mostly perceive implementation as homogenous among firms. Nonetheless, adoption of management standards is heterogeneous as certified firms may not all implement CMSS comprehensively; some firms may fully commit to the requirements of these standards while others might not. Previous studies (Biazzo 2005; Boiral 2003) have shown that, although, some firms do not meet the requirements of a CMSS, could still be able to externally certify their management standard and proclaim the adoption of corporate responsible practices. Thus, adoption of certifiable standards differs from implementation; a distinction not very often addressed in the existing literature (Aravind and Christmann 2011; Gondo and Amis 2013). Acknowledging this difference, however, is of paramount importance as it can assist research into differentiating between firms that apply CMSS as a latest fad and those that fully commit to the CMSS requirements; daily use their underlying practices; and use CMSS as means of improving firms' managerial effectiveness (Nair and Prajogo 2009).

Relatively few studies have until now focused on different levels of CMSS implementation among firms. Specifically, Boiral (2007) found that there are significant differences between CMSS rhetoric and daily practices; firms might ceremonially claim adoption of CMSS so as to comply with institutional forces, whereas in reality they pay lip service to the implementation of such standards. In a similar vein, Yin and Schmeidler (2009) argue that despite isomorphic pressures for adopting CMSS, these standards may be implemented differently in various companies leading to heterogeneous results on the performance of CMSS. The authors do not downplay the importance of CMSS; on the contrary, they suggest that CMSS' effectiveness depends on the level to which companies incorporate the CMSS' documents and practices into their daily operations. The heterogeneous adoption of CMSS is further supported by other studies (Boiral and Amara 2009; Christmann and Taylor 2006), which propose that firms' approach towards CMSS implementation will depend on the costs and benefits stemming from implementation. Additional insights (Heras-Saizarbitoria et al. 2011), stress the need to focus on firms' motives for implementing a management standard as a means of unveiling the true strength of commitment of businesses to CMSS.

Despite valuable contributions provided by existing research there still exists a relative paucity of studies with reference to whether certified firms actually implement the CMSS requirements as prescribed by these standards. Drawing on Christmann and Taylor (2006), we address this issue by analysing what motivates firms to fully conform to the CMSS requirements. We distinguish between firms that use daily the standards' documents and consistently conform to their requirements (substantive implementation) and firms that do not implement the prescribed by CMSS practices (symbolic implementation).

## 2.3 *Motivations*

### 2.3.1 Competitive Motives

Several studies emphasise that firms expect that CMSS will increase their competitiveness (Corporate Watch 2006; Neumayer and Perkins 2005). Firms will weigh the expected benefits against the cost of CMSS, and will only adopt a CMSS if the perceived benefits exceed the costs of this investment. Depending on the strategic goals of each firm, CMSS may help them to increase their competitiveness in two ways: (a) by reducing their costs through efficiency and productivity gains (Goedhuys and Sleuwaegen 2013); and (b) by signalling their increased quality to the markets and thus, differentiating from their competitors (Lannelongue et al. 2013).

### 2.3.2 Cost Competitiveness

In this context, several studies have focused on the competitive benefits associated with adoption of CMSS. Salomone (2008) argues that the most important factors driving the adoption of a standard are internal benefits, such as cost reductions. Equally, Casadesus and Karapetrovic (2005) argue that ISO 9001 offers companies enhanced productivity and by extension improved financial performance whereas Martínez-Costa et al. (2009) and Heras-Saizarbitoria et al. (2011) claim that adoption of CMSS has a positive impact on business performance. Similarly, Boiral (2011) found a significant positive relationship between the drivers for adopting CMSS and competitive benefits.

The few studies that have focused on the heterogeneous implementation of CMSS (Boiral and Amara 2009; Christmann and Taylor 2006), suggest that firms' approach towards CMSS implementation will depend on the costs and benefits stemming from implementation. Firms will fully conform to the CMSS' requirements only when there are internal benefits to be gained from CMSS; otherwise their approach towards management standards will be superficial satisfying only legitimacy purposes and promoting their corporate profile. Existing research (Jamali 2010), pinpoints that firms prefer such an opportunistic behaviour as it enhances managerial discretion and autonomy and enables firms to reap the benefits of adoption without having to bear the costs of implementation.

In sum, recent studies suggest that internal motives, such as improving productivity, profitability and performance, influence the benefits, and by extension the quality of CMSS implementation (Heras-Saizarbitoria et al. 2011). Hence, the first hypothesis states that:

**Hypothesis 1** Firms driven by cost competitiveness motives are more likely to implement CMSS in a substantive way

### 2.3.3 Quality Signalling

Firms may gain further benefits from CMSS by signalling their increased quality to the markets. Kollman and Prakash (2002) conceptualize management standards as club goods, i.e. goods that because their consumption cannot be priced they are provided on the basis of membership fees, and suggest that companies will get certified only when they gain excludable benefits that cannot be gained by competitors. In other words, adoption of CMSS will enable firms to strategically differentiate themselves from their rivals and hence, capture a larger share of the market.

The idea of external signalling as a motivation for adoption of CMSS is also advocated by Terlaak (2007) who uses Spence's signalling model[1] under asymmetric information (1973). Terlaak (2007) proposes that due to information asymmetries in some markets, buyers cannot directly assess the performance of their prospective suppliers; hence, suppliers use CMSS as a signal of superior performance. Specifically, in innovation intensive industries or when technological complexity is high, adoption of CMSS can signal a firm's superior but tacit characteristics (Terlaak and King 2006).

The above literature illustrates that firms will adopt CMSS for signalling their superior quality; in this sense, firms could use certification just as a signal and, hence, implement it symbolically rather than fully commit to the requirements of the CMSS. Christmann and Taylor (2006), however, suggest that firms might behave differently and substantively implement a standard, if the issue addressed by the CMSS is important for their customers (Christmann and Taylor 2006). Therefore, when firms are under pressure from other certified firms or seek to satisfy the requirements of domestic and more importantly international markets they may tend to implement CMSS in a substantive way. This is because these

[1]Spence's argument is that when there is asymmetric information, productive workers will attend a college for differentiating themselves from the non-productive, as long as the college diploma is perceived by stakeholders as a credible signal of good performance capabilities (1973).

firms would like not only to signal their increased quality but also to maintain their brand-name reputation in these markets. In order to do so, they will have to commit to the values and principles of the CMSS in a substantive way. Hence, the second hypothesis states:

**Hypothesis 2** Firms driven by quality signalling motives are more likely to implement CMSS in a substantive way.

### 2.3.4 Institutional Pressures

Existing research highlights the influence of the institutional environment on companies' decision to adopt CMSS (Bartley 2003; Darnall and Edwards Jr 2006; Delmas 2002; Delmas and Toffel 2003; Jiang and Bansal 2003; Waddock et al. 2002; Weaver et al. 1999). These studies suggest that practices supported by the institutional environment are the ones, which are perceived as legitimate forms of behaviour by society and this is why firms end up adopting them (Glynn and Marquis 2004). Companies do not want to deviate from what is perceived as normal behaviour because if they do their legitimacy (Deephouse 1996) and even their survival will be threatened (Brown and Deegan 1998; Deegan and Rankin 1996). In particular, failure to secure legitimacy may bring in the imposition of sanctions, such as fines or boycott of company's products by society. Firms acknowledge the significance of maintaining their legitimacy and for that reason they end up adopting the widely-accepted practices supported by the institutional environment (Garriga and Melé 2004).

However, firms that adopt CMSS for legitimacy purposes may implement it in a symbolic way without fully complying with the standard's requirements. Jiang and Bansal (2003) argue that a CMSS is not adopted by companies for its functional benefits but as an effective means of declaring the company's operation in an acceptable way. In particular, companies use strategies that enable them to manipulate their environment in order to increase their freedom from social control and intervention (Oliver 1997). According to this view, businesses do not passively respond to demands for CSR practices but they employ communication and self-regulatory tools, such as CMSS and codes of ethics, for convincing their stakeholders about the legitimacy of their operations. As Bansal and Hunter (2003) argue although firms can build in house environmental management systems they prefer certified EMS such as ISO14000 because its external evaluation and certification confers legitimacy to the system.

Owing to the fact that adoption of CMSS is mainly motivated by opportunistic rather than responsible behaviour (Campbell 2007; Lenox 2006), companies will tend to implement CMSS mainly in a symbolic manner. Thus, based on this evidence we argue that:

**Hypothesis 3** Firms driven by institutional pressures are more likely to implement CMSS in a symbolic way.

# 3 Methodology

## *3.1 Sample and Data Collection*

This empirical study is based on a survey conducted in Greece. The country provides an interesting research setting for two reasons. First, taking into consideration that Greece has experienced the impact of the 2008 financial crisis severely, it provides a unique empirical context for examining what priority firms assume for CMSS within the current economic climate. Second, most accounts on CMSS come from the USA, Western Europe, and China, and, thus, researchers have called for further empirical investigation in other institutional contexts in order to enrich knowledge on the topic (Heras-Saizarbitoria and Boiral 2013a; Lee 2008; Lindgreen and Swaen 2010).

Primary data was collected through random sampling from companies operating in the county of Attica. We focused on this region because: (a) the county of Attica is the dominant region of Greece in terms of economic activity, accounting for half of the Greek GDP; and (b) a geographically limited sampling frame ensures that all companies share the same cultural and institutional context. Thus, all firms in the sample were likely to face similar levels of scrutiny insofar as they conform to the same social and environmental demands and hence can be analysed alike (Hoffman 2001; Long and Driscoll 2008).

To ensure high quality of responses the key informant method was used (Campbell 1955; Kumar et al. 1993). In particular, the quality managers responsible for the implementation of ISO 9001 in each company were contacted. Previous studies indicate that quality managers are the most knowledgeable and appropriate individuals to complete a questionnaire in ISO 9001 (Christmann and Taylor 2006). Middle-level managers, like quality managers, are also more likely to present a more realistic view of how firms implement CMSS compared to general managers or chief executive officers, who tend to present an ideal image of their company.

Currently, there are more than 4800 ISO 9001 certified companies in Greece (ISO 2012). We identified 80% of these companies in existing databases (ICAP and Quality Net) and after conducting meetings with certification bodies' representatives. Of these, 986 firms (20%) were found to operate in the county of Attica. Following the pilot-testing of the instrument, the survey was fielded for two months and included two reminders. To increase the response rate, respondents were offered the option of answering the questionnaire via a web-site. Additionally, to ensure that respondents will not provide socially biased answers they were not asked to identify themselves or their company; the survey simply had a control number to keep track of respondents versus non-respondents. By the survey's closing date, 201 respondents completed and returned the questionnaire securing a response rate to 20.4%. The latter is above the typical response rate of large-scale mail survey research (De Pelsmacker and Janssens 2007; Dillman 2000).

## 3.2 Testing Data Quality

Non-response bias was tested by comparing early and late respondents (Luo et al. 2009; Pérez-Nordtvedt et al. 2008). T-tests conducted showed no significant differences suggesting sample representativeness and indicating that the survey's sample does not suffer from non-response bias (Aulakh and Gençtürk 2008; Heavey et al. 2009). The reliability of the self-reported data was assessed by randomly contacting through telephone 10 companies chosen from the survey sample (Lazzarini et al. 2008). Information was asked on issues raised in the questionnaire. The information collected was then checked with the responses of the questionnaire. No bias was identified with the data provided by respondents.

We assessed the data for possible bias arising out of common method variance (CMV). CMV is a type of spurious correlation, which occurs amongst constructs when these derive from a common source. It has been shown that the levels of CMV are approximately 30.5% in education, 28.9% in psychology, 15.8% in marketing, and 3.8% in other business areas (Malhotra et al. 2006). Our research is based on the measurement of simple and clear constructs—such as daily use of standards, major non-conformances, cost competitiveness, and institutional motives of CMSS—that we would expect to be associated with lower levels of CMV as it is in the case of 'other business areas'. Yet, we tested the survey for potential presence of CMV, through the use of Harman's (1967) single factor test, and our findings indicate that there are no significant concerns in relation to the CMV problem.

## 3.3 Variable Measurement

This section illustrates the measurement of the variables. The questions that measure the dependent variable—*Quality implementation of ISO 9001*—refer to 2009. The questions related to the measurement of the independent and control variables refer to the 3 year period 2006–2009. This enabled us to avoid endogeneity (Mosakowski 1998) and test the effects of the various motivations (corresponding to the previous 3 years) upon Quality implementation of ISO 9001.

A five-point Likert scale was used to measure the survey items. Factor analysis was conducted for the construction of the independent variables. The reliability of each construct was assessed using Cronbach's alpha; all constructs had $\alpha > 0.7$, thus indicating adequate level of reliability of the survey items (Nunnally 1978).

The discriminant validity of all constructs was assessed by observing the correlation of all independent variables. If the correlation of two variables is very close to 1, then this implies poor discriminant validity (Morrison 1976). Table 1 reports the correlation matrix of all variables; all independent variables meet the discriminant validity test since none of the pair-wise correlations is larger than 0.23 at $p < 0.0010$.

**Table 1** Descriptive statistics and correlation matrix

| Variables | Mean | SD | 1. | 2. | 3. | 4. | 5. | 6. |
|---|---|---|---|---|---|---|---|---|
| 1. Quality implementation of ISO 9001 | 2.01 | 0.69 | 1.000 | | | | | |
| 2. Cost competitiveness | 9.98e−08 | 1 | 0.187*** | 1.000 | | | | |
| 3. Quality signalling | 5.51e−08 | 0.99 | 0.133* | −0.000 | 1.000 | | | |
| 4. Quality signalling_Logo | 1.95 | 1.33 | 0.108 | 0.084 | 0.048 | 1.000 | | |
| 5. Institutional pressures | 2.40e−09 | 1 | −0.064 | −0.000 | 0.000 | 0.100 | 1.000 | |
| 6. Firm size | 3.89 | 1.59 | −0.180** | 0.062 | −0.146** | 0.235*** | −0.043 | 1.000 |
| 7. Profit | 7,115,528 | 4.69e+07 | −0.127* | −0.096 | 0.114 | 0.009 | −0.075 | 0.231*** |

*$p < 0.10$; **$p < 0.05$; ***$p < 0.01$

### 3.4 Dependent Variable

In line with previous empirical research (Christmann and Taylor 2006; Naveh and Marcus 2004), the dependent variable—*Quality implementation of ISO 9001*—was calculated as the average of three indicators that were measured in the survey. These survey items, shown below, were measured in a five-point Likert scale ranging from 1 = not at all, to 5 = to a very large extent. A large score indicates substantive implementation whereas a low score indicates symbolic implementation of ISO 9001.

(a) 'To what extent are the documents created for the purpose of ISO 9001 used in daily practice?' This survey item is understood in the literature to be related to the quality of ISO 9001 implementation (Christmann and Taylor 2006; Naveh and Marcus 2004) since firms that use daily the standard's documents are truly interested in gaining the benefits related to the standard.

(b) 'To what extent is the content of the documents created for the purpose of ISO 9001 changed pending the external audit?' (reverse scored). Previous studies (Christmann and Taylor 2006; Naveh and Marcus 2004) have used this item as an indicator of quality of implementation as firms that change, to a large extent, the content of the documents before the external audit tend to implement the standard in a symbolic way. Specifically, following the initial certification audit, firms that use the standard symbolically, will leave relatively intact the management system, created in the context of ISO 9001, until the next external audit (usually 1 year after). In these cases, a consultant is often called few days before the re-audit to collect all evidence, record and cover all information that should have been collected in the time passed after the last audit. The intention is to mislead the auditors and make them believe that the company has fully complied with the standard's requirement during the time between the two external audits. In contrast, companies that use the management system on their daily operations do not need to call a consultant to change the content of the firm's documents in order to demonstrate compliance with the standard's requirements. Such companies have integrated the standard's documents in their daily operations; they comply with the standard's requirements and do not need any consultants to manipulate the content of the standard's documents.

(c) 'How many major non-conformances does an internal audit reveal on average?' (reverse scored). Major non-conformances refer to major declinations or malfunctions from the requirements of ISO 9001. This indicator is related to the quality of ISO 9001 implementation as the discovery of a major non-conformance effectively signifies that the standard has been defunct and not implemented systematically. As it has been previously mentioned, the implementation of ISO 9001 is externally audited at least once per year. Firms that fully commit to the standard's requirements should not have any major non-conformances (ISO 2009).

## 3.5 Independent Variables

Company managers were also asked to assess in a five-point Likert scale[2] 'the influence of the following twelve factors in your decision to adopt and implement management standards' (see Table 2). We use factor analysis for the purpose of data reduction. Three factors explain approximately 65 percent of the cumulative variance of the twelve variables (eigenvalue > 1). Table 2 presents the three factors and the variables that explain them. The first factor reflects the cost competitive motivations owing to the loading of the following variables: increase in sales, cost savings, greater productivity, and improved financial performance. The second factor captures the quality signalling motivations since it is explained by the variables: pressures from other certified companies, access to international markets, satisfaction of domestic market requirements, and satisfaction of customer requirements. Finally, the third factor reflects the institutional motivations of the company since it is explained by the variables: improved relations with local community, improved relations with governmental authorities, satisfaction of EU requirements, and improved relations with NGOs.

Following relevant arguments in the literature, indicating that firms may use the standard's certification logo as a signal of good practice with the objective of improving their reputation and market acceptance (Jiang and Bansal 2003; Lenox 2006; Terlaak 2007), an additional variable, namely LOGO, was used to measure the quality signalling motivations. Respondents were asked to evaluate in a five-point Likert scale[3] whether they '*use the certification's logo on their products as signal of good product quality*'.

## 3.6 Control Variables

Drawing on evidence found in the literature, indicating that SMEs might have problems with implementing ISO 9001 due to lack of resources (King and Lenox 2000; King et al. 2005), *firm size* was included in the analysis by taking into consideration the logarithmic form of the number of employees. We also controlled for firm performance by incorporating the *profit* variable in the regressions and industry variation by including *industry* dummies.

[2]The Likert scale ranges from: 1 (not influence), 2 (little influence), 3 (some influence), 4 (influential), 5 (very influential).

[3]The Likert scale ranges from: 1 (never), 2 (seldom), 3 (sometimes), 4 (often), and 5 (always).

**Table 2** Principal component analysis

| | Factors | | | | |
|---|---|---|---|---|---|
| | Mean | SD | Cost competitiveness | Quality signalling | Institutional pressures |
| Increase in sales | 2.33 | 1.15 | **0.679** | 0.377 | 0.110 |
| Cost savings | 2.74 | 1.21 | **0.834** | 0.180 | 0.164 |
| Greater productivity | 2.56 | 1.16 | **0.886** | 0.030 | 0.077 |
| Improved financial performance | 2.84 | 1.15 | **0.781** | 0.140 | 0.240 |
| Pressures from other certified companies | 2.80 | 1.29 | 0.001 | **0.527** | 0.291 |
| Access to international markets | 2.28 | 1.31 | 0.137 | **0.539** | 0.333 |
| Satisfaction of domestic market requirements | 1.95 | 0.97 | 0.158 | **0.827** | 0.064 |
| Satisfaction of customer requirements | 1.84 | 0.98 | 0.402 | **0.719** | 0.024 |
| Improved relations with local community | 3.42 | 1.31 | 0.339 | 0.036 | **0.705** |
| Improved relations with governmental authorities | 2.97 | 1.31 | 0.110 | 0.078 | **0.879** |
| Satisfaction of EU requirements | 2.50 | 1.34 | 0.063 | 0.340 | **0.698** |
| Improved relations with NGOs | 3.23 | 1.30 | 0.112 | 0.211 | **0.741** |
| Cronbach's alpha | | | $\alpha = 0.856$ | $\alpha = 0.701$ | $\alpha = 0.807$ |

## 3.7 *Method*

### 3.7.1 Ordinary Least Squares Regression

Ordinary least squares regression was used to examine the motivations of the quality implementation of ISO 9001. To correct for heteroskedasticity, we used the Huber–White standard errors (Greene 2003).

### 3.7.2 Simultaneous Bootstrap Quantile Regression

A small number of studies in the literature look at the factors that influence the quality of CMSS implementation (Christmann and Taylor 2006; Aravind and Christmann 2011; Lannelongue et al. 2013). However, the focus of these studies is on the impact of the various motives upon the average quality of implementation. This is mainly due to the fact that these studies estimate an Ordinary Least Squares regression, which captures how the mean quality implementation changes with the various explanatory variables. Yet, they cannot answer an important question: Do these factors influence quality implementation differently for firms with high levels of quality implementation than for those with average quality implementation? We address this question by conducting a Simultaneous Bootstrap Quantile regression, which allows the effects of the explanatory variables to differ over the average and higher quantiles (Greene 2003). This is mainly due to the fact that we are interested to examine which factors affect not only the average substantive implementation of CMSS but the highly substantive implementation of CMSS as well.

## 4 Results

Table 1 presents the descriptive statistics of the main variables and their correlation. The dependent variable has a mean value of 2.01 and standard deviation of 0.69, which indicates that there is considerable variation in the quality of implementation of ISO 9001 amongst the firms in our sample. Table 3 sheds more light into the distribution of firms with respect to the quality implementation of ISO 9001. The results indicate that adoption of ISO 9001 does not imply substantive implementation; in particular, only 40% of the sampled firms are highly committed into implementing ISO 9001 in a substantive way (above average; >2.01), whereas the majority of the firms (60%) implement ISO 9001 in a symbolic way (below average and average; $\leq 2.01$).

Table 4 presents the regression results for the OLS regression with robust standard errors and the simultaneous bootstrap Quantile regression. The OLS regression reports the marginal effects of the explanatory variables upon the dependent variable—*Quality implementation of ISO 9001*. The Quantile regression

**Table 3** Quality implementation of ISO 9001

| | Below average (<2.01) | Average (=2.01) | Above average (>2.01) | Total |
|---|---|---|---|---|
| Number of firms | 94 | 27 | 80 | 201 |
| Percentage of firms (%) | 47 | 13 | 40 | 100 |

**Table 4** Regression results

| Explanatory variables | Dependent variable: quality implementation of ISO 9001 | | | | |
|---|---|---|---|---|---|
| | OLS regression-robust standard errors | Simultaneous bootstrap quantile regression | | | |
| | Marginal effects (*dy/dx*) | 50th | 70th | 85th | 90th |
| Cost competitiveness | 0.122*** (0.049) | 0.126* (0.066) | 0.078 (0.092) | 0.058 (0.088) | 0.081 (0.082) |
| Quality signalling | 0.079* (0.045) | 0.096 (0.060) | 0.010 (0.097) | 0.115 (0.079) | 0.049 (0.077) |
| Quality signalling_Logo | 0.073** (0.035) | 0.048 (0.050) | 0.141* (0.079) | 0.125** (0.057) | 0.094** (0.046) |
| Institutional pressures | −0.060 (0.048) | −0.004 (0.076) | −0.165** | −0.130* (0.070) | −0.149** (0.072) |
| Constant | 2.144*** (0.182) | 1.756*** (0.256) | 2.325*** (0.359) | 2.984*** (0.358) | 3.471*** (0.342) |
| *Control variables* | | | | | |
| Firm size | −0.085*** (0.030) | −0.091** (0.046) | −0.096* (0.055) | −0.094* (0.048) | −0.113** (0.047) |
| Profit | −1.23e−09 (6.21e−10) | −9.10e −11 (2.71e −09) | −1.23e −09 (2.30e −09) | −2.56e −09 (2.35e −09) | −2.80e −09 (2.38e −09) |
| Industry | 0.034 (0.057) | 0.119 (0.080) | 0.067 (0.100) | −0.038 (0.097) | −0.127 (0.089) |
| $R^2$ | 0.116 | | | | |
| *F* test | 5.41*** | | | | |
| Observations | 201 | | | | |

$^*p < 0.10$; $^{**}p < 0.05$; $^{***}p < 0.01$

reports besides the average (i.e. 50th), the higher quantiles (i.e. 70th, 85th and 90th) of the *Quality implementation of ISO 9001*. As mentioned in Sect. 3.7, we are interested to examine the impact of the explanatory variables not only upon the mean/median of our dependent variable but more importantly upon the higher percentiles of the distribution of *Quality implementation of ISO 9001* so as to determine the factors that lead/deter the higher quality implementation of ISO 9001.

Hypothesis 1 suggests that firms, driven by cost competitiveness motives, are more likely to implement CMSS in a substantive way. This hypothesis is supported by the data analysis. The findings of the OLS regression indicate that cost competitiveness motivates firms to implement CMSS in a substantive way. In particular, the coefficient of the variable—*cost competitiveness*—is positive and statistically significant ($p < 0.05$) in the OLS regression model. However, OLS regression estimates the impact of the explanatory variables upon the mean value of the dependent variable. The Quantile regression allows us to test the impact of the explanatory variable upon the different quantiles of the dependent variable, not only at the mean or median value. The results of the Quantile regression show that while cost competitiveness indeed motivates firms to implement CMSS in a substantive way, this effect holds only among firms that conduct an average quality implementation (50th quantile). The effect is still positive, but not statistically significant at higher quantiles; that is among those firms that are fully committed to the standards' principles (85th and 90th quantiles).

Hypothesis 2 states that firms driven by quality signalling motives are more likely to implement CMSS in a substantive way. This hypothesis is supported by the data analysis. In particular, both relevant coefficients—*quality signalling* and *quality signalling_Logo*—are positive and statistically significant ($p < 0.01$ and $p < 0.05$ respectively). More importantly, the effect of *quality signalling_Logo* is stronger at the higher quantiles (85th and 90th) of the distribution of the dependent variable, which indicates that quality signalling is particularly important in driving the higher quality of ISO 9001 implementation.

Hypothesis 3 suggests that firms driven by institutional pressures are more likely to implement CMSS in a symbolic way. This hypothesis is supported by the analysis of the data. In particular, the coefficient of the variable—*institutional pressures*—has a negative sign but it is not statistically significant, neither at the mean (i.e. in the OLS regression) nor at the median (i.e. in the Quantile regression). Interestingly, the coefficient of the variable *institutional pressures* is negative and statistically significant ($p < 0.05$) at the higher quantiles (70th, 85th and 90th) of the distribution of dependent variable—*Quality implementation of ISO 9001*. This indicates that institutional pressures deter firms from implementing CMSS in a substantive way. Specifically, firms at the higher quantiles (70th, 85th and 90th), which are fully engaged in the implementation of ISO9000, are affected negatively by institutional pressures. This implies that institutional pressures are influential in motivating the symbolic implementation of CMSS by firms only.

## 5 Discussion

The purpose of this chapter was to understand what motivates firms to substantially implement CMSS. The study was driven by the limited quantitative evidence on the topic and by relevant calls in the literature (Iatridis and Kesidou 2016; Heras-Saizarbitoria and Boiral 2013a; Nair and Prajogo 2009) for studies that focus on the quality of implementation of these standards. The latter is critical as information on whether firms fully conform to the CMSS requirements or not, can provide valuable information as to whether firms use these standards as managerial fads or as management practices that can assist them to improve their corporate performance, with respect to the issues addressed by each standard.

This work makes two significant contributions. First, it provides quantitative evidence to a relatively limited body of literature, which analyses quality of implementation of CMSS. In doing so, the chapter offers new insights about the factors that drive firms to substantially implement these standards. Second, it extends previous research on this topic, as it does not only analyse the factors that drive average quality implementation of CMSS but, also, examines the factors that motivate firms to engage in higher levels of implementation of CMSS. Put differently, it focuses not only on average performers but also on higher performers.

The data used were drawn from a survey conducted in Greece and concerned ISO 9001 certified companies. To answer the chapter's question, we reviewed literature related to quality of implementation and firms' motives for engaging into CMSS. In doing so, three hypotheses were developed concerning the relationship between quality of CMSS implementation and cost competitive motives, quality signalling and institutional pressures. The first two hypotheses suggested that cost competitive motives and quality signalling drive the substantive implementation of CMSS whereas the third hypothesis stated that institutional motives relate to symbolic implementation of these standards. The results from our analysis support these three hypotheses.

More precisely, the findings on competitiveness are consistent with previous research (Salomone 2008), which suggests that cost competitiveness is an influential driver of CMSS implementation and corroborate the view that firms' conformance to these standards will be positively associated with financial benefits (Boiral and Amara 2009; Christmann and Taylor 2006; Jamali 2010). The interesting aspect of our results is that cost competitiveness was found to be an influential driver for average performers only. For firms that were fully committed to the standard's requirements, cost competitiveness did not hold any influence.

The findings on quality signalling indicate that quality signalling is an influential driver of substantive implementation, especially for higher performers. Our results support Christmann and Taylor's (2006) view which states that firms, interested in signalling their high quality approach to CMSS, will implement CMSS in a substantive way in order not to threaten their reputation.

The findings on institutional motives are consistent with Heras-Saizarbitoria and Boiral's (2013b) results and lend support to the view that institutional pressures might not be influential as neo-institutional theory suggests. Especially for high performers, our results suggest that institutional pressures deterred substantive implementation of CMSS. This finding does not downplay the significance of legitimacy. Combined with the analysis on quality signalling, these results suggest that firms take legitimacy into account; this is why they are signalling their approach to CMSS in the first place. The analysis mostly indicates that firms do not passively respond to institutional pressures as neo-institutional theory implies. In order to do so, firms want to ensure some benefits; otherwise they will implement CMSS symbolically. This analysis, also, lends support to the idea proposed by Jamali (2010) that when firms do not have anything to gain from, they will go for the symbolic implantation of standards as such approach will enable them to reap the benefits of adoption without having to bear the costs of implementation.

Overall, the chapter's findings on competitiveness and institutional pressures reveal the significant role intrinsic benefits play for the implementation of CMSS. Opposite to the institutional view on CMSS, the fact that 60% of our sampled companies did not comply with the standard's requirements and implemented it symbolically, indicates the high role internal motives are likely to play in the implementation of CMSS.

## 6 Implications for Research

The study's findings have important implications for research and advance knowledge on the quality of implementation of CMSS in several ways. Primarily, by distinguishing between certification and implementation, the study supports the idea that adoption is different from implementation. The more studies adopt this approach the more effectively research will be able to distinguish between firms that adopt CMSS as managerial fads and fashions and those that use these standards to promote a corporate responsible conduct. In this sense, this study provides a robust basis for a new line of inquiry into the analysis of such practices.

Moreover, the results have implications for researchers who use institutional theory as their main theoretical frame of analysis for CMSS. In line with Heras-Saizarbitoria and Boiral (2013b), our analysis suggests that institutional pressures might not be as influential as it was initially considered as firms might actively respond to such pressures by adopting a management standard yet implementing it in a symbolic way. Hence, the belief that institutional pressures define corporate behaviour might be flawed. This limitation is something that researchers need to take seriously into account when dealing with CMSS.

## 7 Implications for Policy Makers

As discussed at the beginning of this chapter, CMSS play a crucial role into the economy as they are treated nowadays as governance mechanisms of promoting corporate responsibility in the areas of the society, environment, and health and safety. CMSS, like ISO 9001, have been extensively used as a means of introducing corporate responsibility practices into many countries (Iatridis et al. 2016; Tencati et al. 2008) whereas major organizations, such as OECD, the European Commission and the United States Council for International Business (European Commission 2003; OECD 2009), have included ISO 9001 in their lists with the most well-known corporate responsibility initiatives. The findings of the survey that 60% of the sampled companies use this standard symbolically, raises concerns of the reliability of ISO 9001 certification and of CMSS in general.[4] These results have important implications for governments, international organisations and industry associations that are keen in using CMSS as governance frameworks for corporations. This is because if these standards are not implemented properly, their role as self-governing mechanisms of corporate social responsibility needs to be questioned.

## 8 Limitations and Conclusion

This paper is not exempt from limitations, which point towards possible directions for future research. First, the paper adopts the motivations of the survey's respondents as the main motivations within the company. Although, in line with previous research, these persons played a major role in the design, development and implementation of CMSS, the data collected was based on their own experiences and were not triangulated by other managers. Future research could collect information on the opinions of different levels managers within each company to provide a more objective view of each company's motivations.

Second, in line with previous studies, we assumed that, because the creation, strcuture, implementation, and audit requirements of IS09001 are similar to ISO 14001, ISO 26000, and SA8000, firms are likely to exhibit similar paterns of behaviour when implementing all of these standards (Christmann and Taylor 2006; Oskarsson and von Malmborg 2005). An interesting extension of this study would be to include firms that have been certified with standards other than ISO 9001. Last but not least, qualitative research in the form of case studies or interviews could be used to provide further insights into reasons for the comprehensive implementation of CMSS.

Despite these limitations, this contribution enhances our understanding of what motivates firms to substantially implement CMSS. Empirical analysis demonstrates that adoption is different from implementation and cost competitiveness and quality signalling are associated with substantive implementation of these standards while institutional pressures relate to symbolic implementation. Most interestingly, the results suggest that cost competitivenes is influential for average performers of CMSS implementation whereas for companies that are fully committed with the standard's requierments the most important factor, driving substnatinve implementation, is quality signalling. What is more, the perception, which holds institutional pressures as influential drivers of corproate behaviour, is challenged by our paper. Our results, not only do not support this view, but indicate that institutional pressures deter high performers of CMSS from implementing these standards substantively.

[4]Recent studies on ISO 14001 (Aravind and Christmann 2011) have reached the same conclusions.

## References

Aravind D, Christmann P (2011) Decoupling of standard implementation from certification: Does quality of ISO 14001 implementation affect facilities' environmental performance? Bus Ethics Q 21(1):73–102

Albuquerque P, Bronnenberg BJ, Corbett CJ (2007) A spatiotemporal analysis of the global diffusion of ISO 9000 and ISO 14000 certification. Manag Sci 53(3):451–468

Aulakh PS, Gençtürk FE (2008) Contract formalization and governance of exporter–importer relationships. J Manage Stud 45(3):457–479

Bartley T (2003) Certifying forests and factories: states, social movements and the rise of private regulation in the apparel and forest products fields. Polit soc 31(3):433–464

Bansal P, Hunter T (2003) Strategic explanations for the early adoption of ISO14001. J Bus Ethics 46(3):289–299

Behnam M, MacLean T (2011) Where is the accountability in international accountability standards? A Decoupling perspective. Bus Ethics Q 21(1):45–72

Biazzo S (2005) The new ISO 9001 and the problem of ceremonial conformity: How have audit methods evolved? Total Qual Manage Bus Excell 16(3):381–399

Boiral O (2003) The certification of corporate conduct: issues and prospects. Int Labour Rev 142 (3):317–340

Boiral O (2007) Corporate greening through ISO14001: a rational myth? Organ Sci 18(1):127–162

Boiral O (2011) Managing with ISO systems: lessons from practice. Long Range Plan 44(3): 197–220

Boiral O, Amara N (2009) Paradoxes of ISO 9000 performance: a configurational approach. Qual Manage J 16(3):36–60

Brown N, Deegan C (1998) The public disclosure of environmental performance information—A dual test of media agenda setting theory and legitimacy theory.Acc Bus Res Winter 21–41

BSI (2013) Health & safety OHSAS 18001. BSI, London

Campbell DT (1955) The informant in quantitative research. Am J Sociol 60(3):339–342

Campbell JL (2007) Why would corporations behave in socially responsible ways? An institutional theory of corporate social responsibility. Acad Manag Rev 32(3):946–967

Casadesus M, Karapetrovic S (2005) Has ISO9000 lost some of its lustre? A longitudinal impact study. Int J Oper Prod Manage 25(6):580–596

Christmann P, Taylor G (2006) Firm self-regulation through international certifiable standards: determinants of symbolic versus substantive implementation. J Int Bus Stud 37(6):863–878

Corporate-Watch (2006) What's wrong with corporate social responsibility?. Corporate-Watch, Oxford

Cragg W (2005) Ethics codes, corporations and the challenge of globalization. Edward Elgar, Cheltenham

Deegan C, Rankin M (1996) Do Australian companies report environmental news objectively? An analysis of environmental disclosures by firms prosecutedsuccessfully by the environmental protection authority. Acc Auditing Accountability J (2):50

Deephouse DL (1996) Does isomorphism legitimate? Acad Manag J 39(4):1024–1039

Delmas MA (2002) The diffusion of environmental management standards in Europe and in the United States: an institutional perspective. Policy Sci 35(1):91–119

Delmas M, Montiel I (2008) The diffusion of voluntary international management standards: responsible care, ISO 9000, and ISO 14001 in the chemical industry.Policy Stud J 36(1):65–93

Delmas MA, Toffel MW (2003) Institutional pressure and environmental management practices, 11th International Conference of the Greening of Industry Network.San Francisco State University, San Francisco: University of Berkeley

De Pelsmacker P, Janssens W (2007) Model for fair trade buying behaviour: the role of perceived quantity and quality of information and of product-specific attitudes. J Bus Ethics 75(4): 361–380

Deming EW (1982) Out of the crisis. MIT Center for Advanced Educational Services, Cambridge

Dillman DA (2000) Mail and internet surveys: the tailored design method, 2nd edn. Wiley, New York
European Commission (2003) Mapping instruments for corporate social responsibility. European Commission, Luxembourg
European Commission (2012) Using standards to support growth, competitiveness and innovation. European Commission, Luxembourg
Glynn MA, Marquis C (2004) When good names go bad: Symbolic illegitimacy in organizations. In: Johnson C (ed) Legitimacy processes in organizations, Elsevier Ltd, Oxford-Greene, pp147–170
Gondo M, Amis J (2013) Variations in practice adoption: the roles of conscious reflection and discourse. Acad Manage Rev 38(2):229–247
Goedhuys M, Sleuwaegen L (2013) The impact of international standards certification on the performance of firms in less developed countries. World Dev 47:87–101
Greene WH (2003) Econometric analysis (5 ed). Prentice Hall, NJ
Heavey C, Simsek Z, Roche F, Kelly A (2009) Decision comprehensiveness and corporate entrepreneurship: the moderating role of managerial uncertainty preferences and environmental dynamism. J Manage Stud 46(8):1289–1314
Heras-Saizarbitoria I, Arana G, Molina-Azorín JF (2011) Do drivers matter for the benefits of ISO 14001? Int J Oper Prod Manage 31(2):192–216
Heras-Saizarbitoria I, Boiral O (2013a) ISO 9001 and ISO 14001: towards a research agenda on management system standards. Int J Manage Rev 15(1):47–65
Heras-Saizarbitoria I, Boiral O (2013b) Symbolic adoption of ISO 9000 in small and medium-sized enterprises: the role of internal contingencies. Int Small Bus J 33(3):299–320
Hoffman AJ (2001) Linking organizational and field-level analyses: the diffusion of corporate environmental practice. Organ Environ 14(2):133–156
Iatridis K, Kesidou E (2016) What drives substantive versus symbolic implementation of ISO 14001 in a time of economic crisis? Insights from Greek manufacturing companies. J Bus Ethics. doi:10.1007/s10551-016-3019-8
Iatridis K, Kuznetsov A, Whyman PB (2016) SMEs and certified management standards: the effect of motives and timing on implementation and commitment. Bus Ethics Q 26(01):67–94
ISO (2009) Management standards. ISO Central Secretariat, Geneva
ISO (2012) The ISO survey. ISO Central Secretariat, Geneva
ISO (2013) Latest ISO survey confirms boost in management systems, vol 24 February 2014
ISO Focus (2014) Crackdown on counterfeiting. ISO Focus January–February, 11–15
Jamali D (2010) MNCs and international accountability standards through an institutional lens: evidence of symbolic conformity or decoupling. J Bus Ethics 95(4):617–640
Jiang RJ, Bansal P (2003) Seeing the need for ISO 14001. J Manage Stud 40(4):1047–1067
King AA, Lenox MJ (2000) Industry self-regulation without sanctions: the chemical industry's responsible care program. Acad Manage J 43(4):698–716
King AA, Lenox MJ, Terlaak A (2005) The strategic use of decentralized institutions: exploring certification with the ISO 14001 management standard. Acad Manage J 48(6):1091–1106
Kollman K, Prakash A (2002) EMS-based environmental regimes as club goods: examining variations in firm-level adoption of ISO 14001 and EMAS in U.K., U.S. and Germany. Policy Sci 35(1):43–67
Kumar N, Stern LW, Anderson JC (1993) Conducting interorganizational research using key informants. Acad Manage J 36(6):1633–1651
Lannelongue G, Gonzalez-Benito O, Gonzalez-Benito J (2013) Environmental motivations: the pathway to complete environmental management. J Bus Ethics 124(1):135–147
Lazzarini S, Claro D, Mesquita L (2008) Buyer-supplier and supplier-supplier alliances: Do they reinforce or undermine one another? J Manage Stud 45(3):561–584
Lee M-DP (2008) A review of the theories of corporate social responsibility: its evolutionary path and the road ahead. Int J Manage Rev 10(1):53–73
Lenox MJ (2006) The role of private decentralized institutions in sustaining industry self-regulation. Organ Sci 17(6):677–690

Lindgreen A, Swaen V (2010) Corporate social responsibility. Int J Manage Rev 12(1):1–7
Long BS, Driscoll C (2008) Codes of ethics and the pursuit of organizational legitimacy: theoretical and empirical contributions. J Bus Ethics 77(2):173–189
Luo Y, Liu Y, Xue J (2009) Relationship investment and channel performance: an analysis of mediating forces. J Manage Stud 46(7):1113–1137
Malhotra NK, Kim SS, Patil A (2006) Common method variance in is research: a comparison of alternative approaches and a reanalysis of past research.Manag Sci 52(12):1865–1883
Martínez-Costa M, Choi TY, Martínez JA, Martínez-Lorente AR (2009) ISO 9000/1994, ISO 9001/2000 and TQM: the performance debate revisited. J Oper Manage 27(6):495–511
Morrison DF (1976) Multivariate statistical methods. McGraw Hill, New York
Mosakowski E (1998) Managerial prescriptions under the resource-based view of strategy: the example of motivational technique. Strateg Manage J 9(12):1169–1182
Nair A, Prajogo D (2009) Internalisation of ISO 9000 standards: the antecedent role of functionalist and institutionalist drivers and performance implications. Int J Prod Res 47: 4545–4568
Naveh E, Marcus AA (2004) When does the ISO 9000 quality assurance standard lead to performance improvement? Assimilation and going beyond. IEEE Trans Eng Manage Sci 51 (3):352–363
Neumayer E, Perkins R (2005) Uneven geographies of organizational practice: explaining the cross-national transfer and diffusion of ISO 9000. Econ Geogr 81(3):237–259
Nunnally, J. C. (1978). Psycometric theory. 2nd edn.McGraw-Hill, New York
OECD (2009) Annual report on the OECD guidelines for multinational enterprises 2008. Employment and industrial relations. OECD Publishing, Paris
Oliver C (1997) Sustainable competitive advantage: combining institutional and resource-based views. Strateg Manage J 18(9):697–713
Oskarsson K, von Malmborg F (2005) Integrated management systems as a corporate response to sustainable development. Corp Soc Responsib Environ Manage 12:121–128
Pérez-Nordtvedt L, Kedia B, Datta D, Rasheed A (2008) Effectiveness and efficiency of cross-border knowledge transfer: an empirical examination. J Manage Stud 45(4):714–744
Rasche A (2009) Toward a model to compare and analyze accountability standards: the case of the UN Global Compact. Corp Soc Responsib Environ Manage 16(4):192–205
Salomone R (2008) Integrated management systems: experiences in Italian organizations. J Clean Prod 16:1786–1806
Tencati A, Russo A, Quaglia V (2008) Unintended consequences of CSR: protectionism and collateral damage in global supply chains: the case of Vietnam. Corp Gov 8(4):518–531
Terlaak A (2007) Satisfying signaling: corporate social strategy and certified management standards, academy of management 2007 annual meeting 1–6. Academy of Management, Philadelphia
Terlaak A, King A (2006) The effect of certification with the ISO 9000 quality management standard: a signaling approach. J Econ Behav Organ 60(4):579–602
Yin H, Schmeidler PJ (2009) Why do standardized ISO 14001 environmental management systems lead to heterogeneous environmental outcomes? Bus Strategy Environ 18(7):469–486

# ISO 14001 Adoption and Environmental Performance: The Case of Manufacturing in Sweden

**Thomas Zobel**

**Abstract** Environmental management systems based on the international standard ISO 14001 have become widespread in companies globally during the last two decades. This chapter aims to contribute knowledge as to which companies are most likely to adopt ISO 14001, environmental frontrunners or poor performers in need of effective approaches. In addition, the chapter aims to assess the impact of ISO 14001 on concrete environmental performance. The context in focus in this study is manufacturing companies in Sweden. The study focuses on changes in performance over a period of 12 years and includes both ISO 14001-certified companies and non-certified companies. The analysis has been carried out using statistical methods for five different environmental areas: air emissions, water emissions, resource use, energy use and waste. In none of these environmental areas have we been able to find any significant differences between certified and non-certified companies regarding performance improvement prior to ISO 14001-certification or the corresponding period in non-certified companies. Certain positive concrete effects as a result of the ISO 14001 adoption have been found but, in general, the standard seems to have limited effect on the improvement of corporate environmental performance. There are many examples in the literature that ISO 14001 has been successful under certain circumstances but at the same time it seems that ISO 14001-certification does not appear to be a guarantee of neither superior environmental performance nor concrete environmental improvements.

**Keywords** ISO 14001 · Environmental management system · Environmental performance · Manufacturing · Sweden · Market signaling

---

T. Zobel (✉)
Division of Business Administration and Industrial Engineering,
Luleå University of Technology, 97187 Luleå, Sweden
e-mail: thomas.zobel@ltu.se



I. Heras-Saizarbitoria (ed.), *ISO 9001, ISO 14001, and New Management Standards*, Measuring Operations Performance, https://doi.org/10.1007/978-3-319-65675-5_3

# 1 Introduction

From the early 1990s and onwards environmental and sustainability issues has become of ever-increasing strategic importance to companies (Meisner Rosen 2001). These issues had previously mostly been driven by compliance but in recent years they are to a greater extent integrated in corporate internal strategy due to increasing stakeholder pressure. This development has caused many companies to turn their attention towards voluntary environmental and sustainability agreements and standards. The voluntary approach that has become the most popular in recent times for companies to effectively and systematically ensure compliance, to develop more environmentally efficient and processes and improve competitiveness, is the environmental management system (EMS) (Rivera-Camino 2003; Gupta and Piero 2003; Poksinska et al. 2003). Such a system "is a set of management processes and procedures that allow an organization to analyze, control and reduce the environmental impact of its operations and services to achieve cost savings, greater efficiency and oversight, and streamlined regulatory compliance" (Pawar and Risetto 2001).

The EMS approach has gained popularity mostly due to the worldwide distribution and commercial success of the international environmental management standard ISO 14001. During the last two decades, more than 300,000 organizations in 171 countries around the world have chosen to certify their EMS (ISO 2017). Certification to this standard follows a similar path to the development of the ISO 9000-series focusing on quality management. In addition to the widespread adoption of ISO 14001, European industrial facilities have to some extent also adopted the Eco-Management and Audit Scheme (EMAS) regulation (about 3000 registered companies), which is partly based on ISO 14001 (EMAS 2015). The popularity of ISO 14001 has helped to spread the standardized management system approach to other areas such as energy management (ISO 50001), occupational health and safety (OHSAS 18001/ISO 45001) and social accountability (SA8000). The commercial success of ISO 14001 has led to extensive research focusing on different phenomena and issues connected to the adoption of EMSs based on the standard (Nawrocka and Parker 2009; Heras-Saizarbitoria and Boiral 2013; Boiral et al. 2017). This chapter is focused on two of those issues: market signaling and concrete environmental effects of system adoption. The geographical context here is Sweden, which is a relevant choice as Sweden is one of the countries in the world that had embraced the ISO 14001 standards as well as other process standards the most (ISO 2017). The research presented in this chapter is a synthesis of several previous individual studies that all have been focusing on ISO 14001 in Swedish manufacturing.

The area of market signaling (including selection effects) has been addressed before but is has not drawn as much attention as many other phenomenon associated with ISO 14001 certification. Evidence is scarce regarding which companies, from an environmental performance perspective, are more likely to adopt ISO 14001.

Is certification a way to show stakeholders that the company already is at the forefront concerning environmental issues? Or is it possibly the poor environmental performers that are in need of effective approaches that are more likely to adopt ISO 14001? A majority of previous studies suggest that ISO 14001 certification indeed could be seen as a legitimate indicator of superior environmental performance (Florida and Davison 2001; Welch et al. 2002; Toffel 2005; Johnstone and Labonne 2009). These studies all confirm the signaling model, originally formulated by Spence (1973), suggesting that strong performers are more likely to certify to ISO 14001 since poor performers have to invest significantly more to get certified. However, there are also studies that cast some doubt on the view that certification serves as a signal (Hamschmidt and Dyllick 2001; King et al. 2005). In this study, we aim to contribute to the answers to the questions above by comparing the change in environmental performance prior to certification in ISO 14001-certified companies and in non-adopting companies.

The environmental impact of ISO 14001 can no longer be regarded as a novel field of research but it is still very much an active field. The debate concerning the environmental effectiveness of ISO 14001 is still open in academia as well as among practitioners due to the fact that studies in this field show divergent results. As a result of this, all the main recent empirical studies (e.g. Gomez and Rodriguez 2011; Nishitani et al. 2012; Testa et al. 2014) and the latest review studies (Nawrocka and Parker 2009; Heras-Saizarbitoria and Boiral 2013; Boiral et al. 2017) suggest further research that might help to explain the previous contrasting conclusions. Commonly these studies attach great importance to methodological weaknesses in previous studies, mainly bias related to performance variables based on managers' perceptions. They are also to a varying degree highlighting issues such as the lack of consensus regarding the measurement of environmental performance, the lack of performance comparisons with non-adopting companies and the fact that a selection effect might be present where better performance precedes ISO 14001 adoption. As a response to the previous studies we have aimed to do assess the environmental impact of ISO 14001 by means of fact-based environmental data covering a wide variety of environmental aspects.

Following this introduction, the next section outlines the research approach, the data set and the statistical methods used in the study. The results of the study are then presented in separate sections, one for each of the two main issues of focus in this chapter. In the beginning of the results sections the aims of this study are reformulated into more specific hypothesis. Due to the fact that research in this field often has been somewhat vague concerning environmental performance and how it should be measured, the result sections are rather detailed. The readers who are more interested in getting an overall picture of the research carried out are advised to go directly to the summary of results at the end of the two result sections. Lastly, the results are discussed, conclusions are drawn and managerial and policy implications are suggested in a final section.

## 2 Methodology

An approach to study a management concept such as an ISO 14001-based EMS needs to focus on changes in environmental performance over time as EMS is a tool for continuous improvement. Furthermore, the studies must include not only companies with EMS but also ones without systems for comparison (Schylander and Zobel 2003). Finally, a study focusing on the impact of the concept needs to take into account the performance of companies prior to the introduction of ISO 14001 to address any possible selection effects where better performance precedes adoption (Heras et al. 2002; Heras-Saizarbitoria et al. 2011).

### *2.1 Research Approach*

In this study, the year 2000 was chosen as the base year and the change in environmental performance over 3-year intervals between 1994 and 2006, i.e. 6 years prior to certification and 6 years after certification, was studied. The year 2000 was chosen as the base year because the peak concerning new ISO 14001-certifications in manufacturing companies occurred in that year in Sweden. Figure 1 shows a schematic illustration of the time periods that form the basis for analysis.

In this study the environmental improvements prior to ISO 14001 adoption has been analyzed over a short period of time (1997–2000) as well as over a longer period (1994–2000). In addition, the change in the rate of improvement after ISO 14001 adoption compared to the improvement before adoption has also been analyzed over a short (1997–2003) and a long (1994–2006) period of time. The shorter periods has the advantage that the production most probably has not

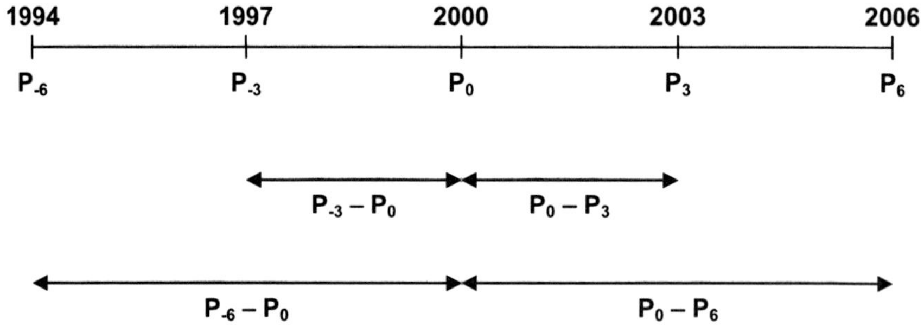


**Fig. 1** Schematic picture of the time periods over which the change in performance has been measured in certified and non-certified companies ($P_{-6}$ = environmental performance in 1994, 6 years before the base year 2000; $P_{-3}$ = environmental performance in 1997, 3 years before the base year 2000; $P_0$ = environmental performance in the base year 2000; $P_3$ = environmental performance in 2003, 3 years after the base year 2000; $P_6$ = environmental performance in 2006, 6 years after the base year 2000)

changed so much during the period and it is therefore easier to follow changes in performance. However it is a disadvantage that changes are most likely smaller than for the longer periods. The major disadvantage with the longer period is that data from 1994 were hard to find, which made the statistical analysis more difficult.

Before analysis, all parameters (e.g. emissions of $SO_2$, waste to landfill or electricity) have been normalized to each company's production rate (e.g. total weight of products or total weight of input raw material), to be able to study changes in environmental performance independent of the level of production. Since the change in environmental performance measured in percent was used, it was possible to use the individual indicators of production rate for each firm as it is given in each firm's environmental permit and reported in its annual environmental report. This means that changes in eco-efficiency have been studied (e.g. ton $SO_2$ ton products$^{-1}$, year$^{-1}$) rather than changes in absolute numbers.

What is actually analyzed in the first sub study (environmental performance before ISO 14001 adoption) is the change in environmental performance over a 3-year and a 6-year period before ISO 14001 certification and the corresponding periods for non-certified companies. The parameters that have been the subject of analysis are therefore the following if the designations from Fig. 1 are used:

$$\textit{Change in environmental performance}\ (\textit{short},\ \%) = ((P_{-3} - P_0)/P_{-3}) * 100$$

$$\textit{Change in environmental performance}\ (\textit{long},\ \%) = ((P_{-6} - P_0)/P_{-6}) * 100$$

Regarding the second sub study (impact of ISO 14001 adoption), the basis for analysis has not been the improvements as such, but rather the change in the rate of improvement before and after ISO 14001 certification and the corresponding periods for non-certified companies. The parameters that have been the subject of analysis have been calculated using the following equations, in which designations from Fig. 1 are used:

$$\begin{aligned}&\textit{Change in rate of improvement}\ (\textit{short},\ \%)\\&\quad = ((P_0 - P_3)/P_0) * 100 - ((P_{-3} - P_0)/P_{-3}) * 100\end{aligned}$$

$$\begin{aligned}&\textit{Change in rate of improvement}\ (\textit{long},\ \%)\\&\quad = ((P_0 - P_6)/P_0) * 100 - ((P_{-6} - P_0)/P_{-6}) * 100\end{aligned}$$

## 2.2 Collection of Data

Mandatory annual environmental reports from companies with regulatory environmental permits have been the basis for objective environmental data used in this study. The reports include detailed records of a large number of environmental aspects that are specific to each company. In addition, they also report on the level of yearly production, typically expressed as number of products, tons produced or tons of certain resources used in production.

Manufacturing companies (European industrial activity classification code 10–33) were chosen for the study because we expected to find relatively complex and significant environmental impact in them. The total number of such companies in Sweden was 2331 at the time of this study. The list was supplemented with information on ISO 14001 certification from all certification bodies in Sweden. A total of 130 manufacturing companies that received their ISO 14001 certificate in 2000 could be found. A random sample of 130 companies was taken from the total population of non-certified manufacturing companies with a permit. Environmental reports for the years 1994, 1997, 2000, 2003 and 2006 were collected from the companies' environmental monitoring agencies. Documents were received for 217 of the 260 companies (83%). Not all the 217 companies were included in the study for various reasons (e.g. insufficient or poor-quality documentation). After the non-usable companies had been excluded, 66 certified companies and 50 non-certified companies remained.

### 2.3 Statistical Methods

Certified and non-certified companies were compared using the $t$ test with a confidence level of 95% (The test gives a $P$ value and if this value is lower than 0.05 there is a significant difference between groups). This is the preferred test for comparing two samples when data can be considered normally distributed and variances are equal (Bethia et al. 1995; Box et al. 2005). The sample distribution has been tested using the parameters standardized skewness and standardized kurtosis (Thode 2002). These two concepts are descriptors of the shape of a probability distribution. Kurtosis is a measure of the "peakedness" of the distribution whereas skewness is a measure of the asymmetry. Variances were tested using an $F$ test, which is a commonly used test to compare variances between groups (Lomax 2007). On several occasions we have used the standard method included in the software IBM SPSS Statistics 22 to identify extreme values (outliers) in the data set.

## 3 ISO 14001: A Tool for Companies in Need?

A comparative statistical analysis of the change of environmental performance between certified and non-certified companies was conducted and the following null hypothesis and alternative hypothesis was tested for the shorter period 1997–2000 and the slightly longer period 1994–2000:

**$H_0$** the change in environmental performance prior to certification for ISO 14001-certified companies does not differ significantly from the corresponding change for non-certified companies.

**$H_1$** the change in environmental performance prior to certification for ISO 14001-certified companies differs significantly from the corresponding change for non-certified companies.

The change in performance was studied in five specific environmental areas: air emissions, water emissions, resource use, energy use and waste. The results in each area are reported in subsections below. The main reason for the division into different areas is to potentially be able to see variances in results concerning different environmental aspects.

## 3.1 Air Emissions

When analyzing the changes concerning air emissions, all emission parameters were merged into one parameter by calculating the average for the percentage change for all air emissions. Typical air emissions are solvents (VOCs), dust, nitrogen oxides ($NO_X$) and sulfur dioxide ($SO_2$), but several other parameters were presented in the reports as well. Parameters that are only expressed as concentrations, e.g. mg/l or mg/m$^3$, have not been normalized to the rate of production since these values are independent of how much is produced.

About half of the companies have parameters so that they can be included in the analysis of the shorter time period, which reduces the possibility to see small differences in a comparative analysis. In Table 1, we see that air emissions are reduced by 22% for the certified companies, while they are reduced by 17% for the non-certified companies. The hypothesis were tested by a $t$ test giving the value $P = 0.673$ and the rather large value tells us that we cannot reject the null hypothesis. Fewer companies reported reliable data for air emissions during the period 1994–2000 compared with the shorter period. In Table 1 we see that air emissions are reduced by 20% for the certified companies, while they are reduced by 4% for the non-certified companies. A $t$ test gives the value $P = 0.431$, meaning that we cannot reject the null hypothesis. It should be noted that there is a statistically significant difference between variances in this case.

## 3.2 Water Emissions

Water emissions have been analyzed in the same way as air emissions. Common types of emissions include heavy metals, BOD, suspended solids, nitrogen, phosphorus, but several other emissions types are reported. As far as it has been

**Table 1** Summary statistics on the change in environmental performance in the periods 1997–2000 and 1994–2000

| Parameters | Certified 1997–2000 | Non-certified 1997–2000 | Certified 1994–2000 | Non-certified 1994–2000 |
|---|---|---|---|---|
| *Air emissions* | | | | |
| No. of companies | 34 | 22 | 30 | 15 |
| Change (%) | 22.0 | 17.2 | 20.3 | 3.59 |
| Standard deviation | 38.9 | 45.2 | 55.3 | 85.0 |
| *Water emissions* | | | | |
| No. of companies | 31 | 26 | 28 | 14 |
| Change (%) | 3.51[a] | 2.34 | −20.9[a] | −3.95 |
| Standard deviation | 48.3 | 37.9 | 65.3 | 95.6 |
| *Resource use* | | | | |
| No. of companies | 32 | 23 | 21 | 16 |
| Change (%) | −0.788 | 6.57[a] | 6.77 | −4.11 |
| Standard deviation | 29.2 | 31.3 | 35.8 | 48.0 |
| *Energy use* | | | | |
| No. of companies | 48 | 23 | 42 | 17 |
| Change (%) | 1.91[a] | 12.7 | 8.78 | 10.6 |
| Standard deviation | 31.5 | 30.9 | 36.1 | 33.2 |
| *Waste* | | | | |
| No. of companies | 51 | 45 | 40 | 20 |
| Change (%) | −6.99[a] | −35.0[a] | −13.8[a] | 7.68 |
| Standard deviation | 64.4 | 105 | 82.9 | 53.4 |

[a]The test using the parameters standardized skewness and standardized kurtosis could not confirm normally distributed data

possible, total emissions over the whole year are used in the analysis but, when this has not been possible, concentrations have been used instead.

As with air emissions, about half of the participating companies have reported some type of emissions to water that could be used in the analysis of the shorter period. The water emissions for the certified companies are reduced by 4%, while they are reduced by 2% for the non-certified companies. A *t* test gives the value $P = 0.920$ and we can therefore not reject the null hypothesis. The material collected in this study gives us relatively few useful observations of water emissions for the longer period 1994–2000. The change in performance regarding water emissions for both certified and non-certified companies is negative (−21% of the certified companies and −4% for non-certified companies). The magnitude of the difference might indicate that we indeed have a difference, but a *t* test gives the value $P = 0.502$, which makes us confident that we cannot reject the null hypothesis.

## 3.3 Resource Use

The change in resource use has been difficult to measure by using environmental reports since different types of resources are rarely reported. However, three parameters are relatively often included in the reports: chemical use, water use and the use of fuels (mainly diesel oil) for internal transports. The mean value of the change for these three parameters has been calculated and used in the analysis within the area of resource use. It is not obvious that an increase in chemical use is always worse from an environmental perspective, since it is often the properties of the chemical and not its amount that will determine its environmental impact. In this study, we have disregarded this and made the assumption that a reduction of the amount of chemical use in general is environmentally sound.

About half of the companies in the samples report information concerning one or more of the three types of resources that are used for the analysis of the shorter period. We can see in Table 1 that the resource use has increased by 1% for the certified companies and that the corresponding value for the non-certified companies is a reduction of 7%. A *t* test gives the value $P = 0.375$, which does not give us any reason to reject the null hypothesis. Only about one third of the companies in the samples report one or more parameters within the resource use area over the period 1994–2000. Resource use is reduced by 7% for the certified companies but is increased by 4% for the non-certified companies. Despite the small count of observations, a *t* test has been carried out giving the value $P = 0.434$. The null hypothesis therefore appears to be valid.

## 3.4 Energy Use

When analyzing the change in energy use, three different parameters have been used: electricity, fossil fuel (excluding fuels for internal transports) and total energy. The values that have been the subject of analysis comprise of a merger of the three parameters where the mean value of the percentage changes for the three parameters has been calculated. The parameter fossil fuel use often consists only of consumption of fuel oil but can sometimes include several fossil fuels such as oil, LPG and natural gas. In such cases, the total energy content of the various fuels has been calculated. Parameters that are related to renewable energy sources are not included as separate parameters because an increase in the use of renewable energy sources can be counted either as an improvement (if fossil fuels are replaced) or as deterioration (if the total energy use increases). Renewable energy is therefore hard to use as a parameter in this study.

Compared with the previously analyzed areas, significantly more companies (about two thirds) report of one or more of the parameters we can use for the shorter period in the area of energy use. The certified companies have reduced their energy use by 2% while the non-certified companies perform slightly better with a reduction by 13%. As before, variances are large. Testing of the hypothesis by a *t* test gives the value $P = 0.177$. Thus, we cannot reject the null hypothesis. A relatively small number of non-certified companies reported energy-related parameters in 1994, which makes a comparison between the groups difficult for the longer period. We see in Table 1 that the mean value of the change in energy use indicates a similar reduction for both certified (9%) and non-certified companies (11%). A *t* test gives the value $P = 0.857$, and we are therefore not close to rejecting the null hypothesis.

## 3.5 Waste

Waste-related parameters have been handled in a similar way as energy-related parameters. The parameters hazardous waste, waste to landfill and the total amount of waste have been used for analysis. The values that have been the subject of analysis are the combination of these three parameters where the mean values of the percentage change for the three parameters have been calculated. An increase in waste going to reuse, recycling, composting or incineration has not been used for analysis for the same reason as for renewable energy sources.

Despite problems in interpreting environmental reports relating to the disposal of waste and quantities of waste, the number of observations that we have been able to use for the analysis is greater than for the other areas. We see in Table 1 that the certified companies have increased their waste production by 7% but the non-certified companies perform even worse with an increase by 35%. We might suspect that there is a difference between the companies, but a *t* test does not confirm such a suspicion ($P = 0.113$). It should be noted that the *t* test is not fully valid in this case as there is a statistical difference between the variances in the two samples of companies. The quality of the waste-related data described in the environmental reports from 1994 is much poorer than it is in more recent years. Hence, significantly fewer observations are available to use for analysis for the period 1994–2000 than for the shorter period. Waste production for the certified companies was reduced by 14% while the non-certified companies increased their production of waste by 8%. These results indicate a difference between the two groups of companies, but as before the variances are large and we cannot reject the null hypothesis ($P = 0.300$). The validity of the *t* test is questionable since there is a statistically significant difference between the variances.

### 3.6 Summary of Results

To summarize the results of our study concerning the situation before ISO 14001, we have not in any of the analyzed environmental areas been able to find any statistically significant differences between certified and non-certified companies regarding the change in environmental performance prior to ISO 14001-certification or the corresponding period in non-certified companies. Hence, we cannot confirm the proposed signaling model.

## 4 The Environmental Impact of ISO 14001 Adoption

As in the case of the study of the situation before ISO 14001 adoption the study of the possible impact of the standard has been divided into different environmental areas. However, the areas energy use and waste production has been more finely divided into subareas. Otherwise the analyses have been conducted in the same way as the ones concerning the situation prior to certification. The following null hypothesis and alternative hypothesis were tested for the periods 1997–2003 and 1994–2006:

$\mathbf{H_0}$ the change of the environmental performance improvement rate for ISO 14001-certified companies does not differ significantly from the corresponding change for non-certified companies.

$\mathbf{H_1}$ the change of the environmental performance improvement rate for ISO 14001-cerfified companies differs significantly from the corresponding change for non-certified companies.

### 4.1 Air Emissions

About half of the companies have parameters so that they can be included in the analysis of the period 1997–2003, which reduces the chance of detecting small differences in a comparative analysis. As can be seen in the Table 2, the mean rate of improvement is reduced by 43% for the certified companies, while it increases by about 6% for the non-certified companies. A $t$ test gave the value $P = 0.139$ and since the $P$ value is larger than 0.05, we cannot reject the null hypothesis. It should be noted that there is a statistically significant difference between variances in this case. Fewer companies reported reliable data for air emissions during the period 1994–2006 compared with the shorter period. In Table 2, it can be seen that the mean rate of improvement decreases by 26% for the certified companies, while it increases by 15% for the non-certified companies. However, as for the shorter period the null hypothesis cannot be rejected as a *t* test gave the value $P = 0.142$.

## 4.2 Water Emissions

As with air emissions, about half of the participating companies have reported some type of emission to water that could be used for analysis of the shorter period. The average improvement rate for the certified companies decreases by 5%, while it increases by 13% for the non-certified companies. The results of a *t* test indicate a *P* value of 0.443 and the null hypothesis can therefore not be rejected. The material collected in this study included relatively few useful observations of water emissions for the period 1994–2006. It can therefore be expected that it may be difficult to see any differences between the two company groups. The mean change in improvement rate for both certified and non-certified companies is positive (about 10% of the certified companies, and about 5% for non-certified companies) and they are close to each other. The *t* test confirms that it is fairly certain that the null hypothesis cannot be rejected ($P = 0.854$). However, the validity of the *t* test is questionable since there is a statistically significant difference between the average values of the samples' variances.

## 4.3 Resource Use

About half of the companies in the samples report information concerning one or more of the three types of resources that are used in the analysis of the shorter period. The mean rate of improvement increases by 11% for the certified companies and the corresponding value for the non-certified companies is 1% (see Table 2). A *t* test gave the value $P = 0.483$, which does not give any reason to reject the null hypothesis. Only about a third of companies in the samples report one or more parameters within the resource use area during the period 1994–2006. The mean value for the improvement rate increases a few percent for both the certified and the non-certified companies. Despite the small count of observations, a *t* test has been carried out and the test gave the value $P = 0.922$. The null hypothesis therefore appears to be valid.

## 4.4 Energy Use

When analyzing the change in improvements of energy efficiency, two different parameters were used: electricity and fossil fuel consumption (total energy content of oil, LPG and natural gas).

**Table 2** Summary statistics on the change in the rate of environmental improvement between the periods 1997–2000 and 2000–2003 and the periods 1994–2000 and 2000–2006

| Parameters | Certified 1997–2003 | Non-certified 1997–2003 | Certified 1994–2006 | Non-certified 1994–2006 |
|---|---|---|---|---|
| *Air emissions* | | | | |
| No. of companies | 35 | 23 | 29 | 15 |
| Change (%) | −43.1 | 6.01 | −26.1 | 15.0 |
| Standard deviation | 146.9 | 66.8 | 82.7 | 93.0 |
| *Water emissions* | | | | |
| No. of companies | 30 | 25 | 27 | 14 |
| Change (%) | −5.44 | 13.1[a] | 10.3 | 4.52 |
| Standard deviation | 80.5 | 97.1 | 66.0 | 135 |
| *Resource use* | | | | |
| No. of companies | 31 | 23 | 20 | 15 |
| Change (%) | 11.2[a] | 1.28 | 3.52 | 5.29 |
| Standard deviation | 49.8 | 52.2 | 50.2 | 55.3 |
| *Electricity use* | | | | |
| No. of companies | 33 | 12 | 23 | 7 |
| Change (%) | 1.53 | 4.05 | 2.03[a] | 2.46 |
| Standard deviation | 28.2 | 53.9 | 37.5 | 23.7 |
| *Fossil fuel use* | | | | |
| No. of companies | 40 | 22 | 36 | 14 |
| Change (%) | 7.13 | −38.1 | 3.32 | −34.2 |
| Standard deviation | 75.2 | 61.2 | 84.5 | 88.9 |
| *Hazardous waste* | | | | |
| No. of companies | 44 | 37 | 33 | 15 |
| Change (%) | 5.98 | 39.4 | 4.76 | −42.4 |
| Standard deviation | 84.6 | 167 | 130 | 139 |
| *Waste to landfill* | | | | |
| No. of companies | 17 | 14 | 10 | 6 |

(continued)

**Table 2** (continued)

| Parameters | Certified 1997–2003 | Non-certified 1997–2003 | Certified 1994–2006 | Non-certified 1994–2006 |
|---|---|---|---|---|
| Change (%) | −9.50 | 12.4 | 3.30 | 26.2 |
| Standard deviation | 41.3 | 67.0 | 43.7 | 31.5 |
| *Total amount of waste* | | | | |
| No. of companies | 18 | 6 | 10 | 2 |
| Change (%) | −6.89 | −34.3 | −32.8 | −62.0 |
| Standard deviation | 33.8 | 52.8 | 56.4 | 83.4 |

[a]The test using the parameters standardized skewness and standardized kurtosis could not confirm normally distributed data

Only a relatively small number of non-certified companies reported electricity use in 1997 (see Table 2), but we are still able to perform a *t* test with enough validity for a comparison between the two groups of companies since a common rule of thumb is that one should at least have more than ten observations in a sample (Bethia et al. 1995). We see that the mean value of the change in the rate of improvement indicates a slight increase in efficiency for both certified (1.53%) and non-certified companies (4.05%). A *t* test gives the value $P = 0.839$ and we can be rather certain that there is no difference between the two groups of companies. In the same regard as for the shorter time period, non-usable environmental reports for the time period 1994–2006 has a negative effect of the possibility to identify differences between the two groups of companies. However, the average changes in both groups are limited in their extent and the change seems to be very similar in both groups (2.03% in the certified group and 2.46% in the non-certified group). A *t* test gives us no reason to question the null hypothesis ($P = 0.977$).

Almost twice as many of the non-certified companies have reported their use of fossil fuel compared to the use of electricity and a *t* test therefore should have slightly more explanatory power. The poor fossil fuel efficiency in the non-certified companies after the base year compared to the previous period (Average change −34.2%) gives us cause to suspect that the certified companies are performing better. Indeed, a *t* test gives us $P = 0.019$ which makes the difference significant. The low *P* value means that we can be relatively certain that the certified companies are performing better regarding their work towards improved fossil fuel efficiency in production for the shorter time period close to the ISO 14001 certification year. As before, we are also somewhat limited by a low count of observations regarding fossil fuel use over the longer time period due foremost to the quality of the 1994 environmental reports from the non-certified companies. A *t* test gives us $P = 0.171$ which does not make the difference significant even though the non-certified companies seem to perform rather poorly after the base year. However, we have reason to suspect that the certified companies are also performing better than the non-certified regarding fossil fuel use in the longer time period.

## 4.5 Waste

The data collected from the environmental reports allowed us to study three different waste parameters: hazardous waste, waste to landfill and the total amounts of waste (excluding hazardous waste).

Despite some difficulties in interpreting environmental reports relating to the disposal of waste, we have been able to use data from about two-thirds of the certified companies as well as the non-certified companies when it comes to hazardous waste. In Table 2 we see that the average change in hazardous waste generation improvement is 6% for the certified companies and 39% for non-certified companies. However, a $t$ test gives the value $P = 0.248$, which means we are not close to being able to say that there is a significant difference between the groups. It should be noted that there is a significant difference between variances in the two groups. The quality of the reports strongly influences the usable number of hazardous waste observations for the longer period, but we have enough data for a comparison. The mean value of the change in the rate of improvement for the certified companies is 5% while the non-certified companies seem to have performed much worse in the period after year 2000 than they did the previous period (−42%). However, the difference is not statistically significant since a $t$ test gives $P = 0259$.

Only about a third of the companies in the sample report usable data concerning waste to landfill for the shorter period. The average change in waste to landfill generation improvement is almost −10% for the certified companies while the non-certified companies seem to improve their handling of waste to landfill by 12%. However, the $t$ test do not support that any differences would exist between the groups of companies ($P = 0.272$). Concerning waste to landfill (as well as total amounts of waste), data from 1994 is so rare that it is not possible to conduct meaningful statistical analyses for the 12 year period.

A small number of non-certified companies (6 companies) reported reliable data concerning the total amounts of waste in the shorter period (see Table 2, which makes a comparison between the groups somewhat problematic. The mean value of the change in the rate of improvement indicates a slight decrease for certified companies (7%) and what might be a larger reduction for the non-certified companies (34%). We might suspect that the certified companies perform better but a $t$ test does not confirm this ($P = 0.150$).

### 4.6 *Summary of Results*

To summarize the results of our study concerning the environmental impact of ISO 14001, we have found that the standard seems useful concerning energy use but not in the other environmental areas. We have found some evidence that companies adopting ISO 14001 might improve their energy efficiency concerning fossil fuel use at a faster rate than companies choosing not to certify. In contrast, we were not able to find any significant differences concerning electricity use. In other environmental areas such as air emissions, water emissions, resource use and waste production, ISO 14001 does not generally seem to have a positive impact. In fact, it is even so that we can suspect that the non-certified companies are improving more than the certified companies concerning air emissions (even if the difference is not statistically significant).

## 5 Discussion and Conclusions

The results in this study concerning the environmental performance in companies before ISO 14001-certification can be placed somewhat in-between results in previous studies. We cannot confirm empirical results supporting the signaling model (Florida and Davison 2001; Welch et al. 2002; Toffel 2005; Johnstone and Labonne 2009), and we cannot find support to findings indicating that it is the companies that need it the most, that choose to certify to ISO 14001 (Hamschmidt and Dyllick 2001; King et al. 2005). In a sense, the mixed results in this area are comforting since it seems that not only the already environmentally aware companies choose to certify towards ISO 14001 to signal an already superior environmental performance. It seems equally likely that companies with poor environmental performance choose to certify their EMSs. The initial idea of the ISO 14001standard was to be an effective tool to be used by companies and other organizations so that they could contribute to sustainable development. In order to make a substantial contribution, it is necessary that it is not only the environmental frontrunner companies that adopt an EMS and seek certification, but also the companies with poor environmental performance since the environmental improvement potential is far greater in these companies. However, this view is not shared by all scholars. It seems that some would prefer if we could confirm the signaling model so that an ISO 14001 certificate could be used as a market signal for supply chain partners or an environmental compliance signal for public agencies (King et al. 2005; Terlaak 2007).

The findings concerning the concrete environmental impact of ISO 14001 adoption must be regarded as rather disappointing. Some positive effects on energy efficiency has been identified but in general the environmental performance seem to be inert to ISO 14001 certification. With regard to air emissions, it can even be suspected that the non-certified companies are increasing their rate of improvement at a greater rate than the certified companies. A possible reason for this could be that the certified companies simply do not consider air emissions to be a significant environmental aspect in accordance to the specifications in ISO 14001 and therefore do not actively work to improve in this area. In addition, these types of emissions are normally controlled by curtain purification techniques and it might not be easy to continually improve these aspects in the context of EMS improvement programs. Concerning energy, we have found some evidence that certified companies might improve their energy efficiency with reference to fossil fuel use at a faster rate than companies choosing not to certify. In contrast, we were not able to find any significant differences concerning electricity use. The differences in the results are not easily explained but it might be the case that the certified companies estimate that the use of fossil fuel in production is a more important environmental issue than energy efficiency related to the use of electricity and therefore focus their improvement effort on fossil fuels. Such a choice might be natural since fossil fuel use is a rather high profile issue and the electricity production mix in Sweden is not associated with a high degree of $CO_2$ emissions as it originates almost exclusively from nuclear and hydro power.

A number of studies have claimed to study the relationship between ISO 14001 and concrete environmental performance by using fact-based environmental data. However, there is only one study that has analyzed such a large amount of parameters so that a fair comparison can be made with the study presented in this chapter. Generally it was found in this study that ISO 14001-certified companies in several European countries did not perform better than non-certified companies (Hertin et al. 2008). In only one case could statistically significant differences be found between the certified and non-certified companies. In addition, the researchers also found that the certified companies in several cases appeared to perform worse than non-certified companies. Even though the approaches and methods used in this study were different from the ones in the study presented here, the results can be considered to be relatively consistent. Concerning the use of resources, energy and waste, there are not really any other studies to compare with since the methodology differs too much to make a meaningful comparison. But regarding different types of environmentally harmful emissions, this study positions itself closer to the more critical studies based on the American Toxic Release Inventory (TRI) by Matthews (2001), Ghisellini and Thurston (2005) and Darnall and Sides (2008) as well as the Spanish study by Gomez and Rodriguez (2011). It can also be concluded that this study shows mixed results similar to two other TRI-studies (King et al. 2005; Toffel 2005) and a Canadian study on water emissions (Barla 2007).

Our rather mixed results concerning the possible environmental benefits of ISO 14001-adoption provide important input for companies already certified as well as for those considering certification. It seems as if a positive outcome from system adoption and third party certification is no guarantee for concrete environmental improvement even though the literature shown that there are many examples in which the systems have been successful (Boiral et al. 2017). In other words, an EMS in accordance to ISO 14001 can be an appropriate approach for companies wanting to improve their environmental performance but they have to do more than just focus on the certificate. Managers in certified companies and companies that want to or are externally pressured (e.g. by other companies in their supply chain) to implement an EMS should be aware that they have to give their support and sufficient resources to system development and integration in operations. The findings in this study also provide input for the ISO organization issuing the management system standards. However, it looks as though the ISO organization has successfully identified and discussed findings in this research field as the new version of ISO 14001, introduced at the end of 2015, has more focus on management by environmental objectives as well as specifications concerning the assessment of environmental performance (ISO 2015). Hopefully, the same switch in focus will occur at actual certification audits since it seems as if these have been more focused on procedural rather than substantive aspects of the management systems (Heras-Saizarbitoria et al. 2013).

The results of the project also have policy implications. Policy makers should take the finding in this study and in other studies in this field into consideration when they embed voluntary management systems such as ISO 14001 into State implemented environmental improvement programs or environmental legislation. Since ISO 14001-certification does not appear to be a guarantee of neither superior environmental performance nor concrete environmental improvements, third party certification should be supplemented with other control measures before certified companies receive benefits such as exemptions from regulatory requirements or tax credits.

## References

Barla P (2007) ISO 14001 certification and environmental performance in Quebec's pulp and paper Industry. J Environ Econ Manage 53:291–306

Bethia RM, Duran BS, Boullion TI (1995) Statistical methods for engineers and scientists, 3rd edn. Marcel Dekker, New York

Boiral O, Guillaumie L, Heras-Saizarbitoria I, Valery Tayo Tene C (2017) Adoption and outcomes of ISO 14001: a systematic review. Int J Manage Rev 1–22 (in press)

Box GEP, Hunter WG, Hunter JS (2005) Statistics for experimenters: design, discovery and innovation. Wiley-Interscience, Hoboken

Darnall N, Sides S (2008) Assessing the performance of voluntary environmental programs: does certification matter? Policy Stud J 36:95–117

European Commission (2015) Eco-management and audit scheme: reports and statistics. http://ec.europa.eu/environment/emas/register/reports/reports.do;jsessionid=X9whrE8BojI6ZhwV_0dQCgfuqWANZqXLCVDDT9n1eoslo7NwQ-bV!-1719577194. Accessed 12 Aug 2015

Florida R, Davison D (2001) Gaining from green management: environmental management systems inside and outside the factory. Calif Manage Rev 43:64–84

Ghisellini A, Thurston DL (2005) Decision traps in ISO 14001 implementation process: case study results from Illinois certified companies. J Clean Prod 13:763–777

Gomez A, Rodriguez MA (2011) The effect of ISO 14001 certification on toxic emissions: an analysis of industrial facilities in the North of Spain. J Clean Prod 19:1091–1095

Gupta M, Piero T (2003) Environmental business is good business. Ind Manage 45(5):14–19

Hamschmidt J, Dyllick T (2001) ISO 14001—profitable? Yes! But is it ecoeffective? Greener Manage Int 34:43–53
Heras-Saizarbitoria I, Boiral O (2013) ISO 9001 and ISO 14001: towards a research agenda on management system standards. Int J Manage Rev 1:47–65
Heras I, Dick GPM, Casadesús M (2002) ISO 9000 registration's impact on sales and profitability: a longitudinal analysis of performance before and after accreditation. Int J Qual Reliab Manage 19(6):774–791
Heras-Saizarbitoria I, Dogui K, Boiral O (2013) Shredding light on ISO 14001 certification audits. J Clean Prod 51:88–98
Heras-Saizarbitoria I, Molina-Azorìn JF, Dick GPM (2011) ISO 14001 certification and financial performance. J Clean Prod 19:1–12
Hertin J, Berkhout F, Wagner M, Tyteca D (2008) Are EMS environmental effective? The link between environmental management systems and environmental performance in European companies. J Environ Plan Manage 51:259–283
International Organization for Standardization (2017) ISO survey 2015. https://www.iso.org/the-iso-survey.html. Accessed 18 June 2017
International Organization for Standardization (2015) ISO 14001:2015 environmental management systems—requirements with guidance for use. Geneva, Switzerland
Johnstone N, Labonne J (2009) Why do manufacturing facilities introduce environmental management systems? Improving and/or signaling performance. Ecol Econ 68:719–730
King AA, Lenox MJ, Terlaak A (2005) The strategic use of decentralized institutions: Exploring certification with the ISO 14001 management standard. Acad Manage J 48:1091–1106
Lomax RG (2007) Statistical concepts: a second course. L. Erlbaum Associates, Mahwah
Matthews DH (2001) Assessment and design of industrial environmental management systems. PhD dissertation, Carnegie Mellon University
Meisner Rosen C (2001) Environmental strategy and competitive advantage. Calif Manage Rev 43 (3):8–15
Nawrocka D, Parker T (2009) Finding the connection: environmental management systems and environmental performance. J Clean Prod 17:601–607
Nishitani K, Kaneko S, Fujii H, Komatsu S (2012) Are firms' voluntary environmental management activities beneficial for the environment and business? An empirical study focusing on Japanese manufacturing firms. J Environ Manage 105:121–130
Pawar WM, Risetto C (2001) A tool for improvement: environmental management systems. Public Manage 83:10–17
Poksinska B, Dahlgaard JJ, Eklund J (2003) Implementing ISO 14000 in Sweden: motives, benefits and comparisons with ISO 9000. Int J Qual Reliab Manage 20(5):585–606
Rivera-Camino J (2003) What motivates European firms to adopt environmental management systems? Eco Manage Audit 8(3):134–143
Schylander E, Zobel T (2003) Environmental effects of EMS: Evaluation of the evidence. In: Proceedings of evaluation of sustainability European conference—EASY-ECO 2, Vienna, Austria, May 15–17
Spence AM (1973) Job market signaling. Q J Econ 87:355–374
Terlaak AK (2007) Order without law? The role of certified management standards in shaping socially desired firm behaviors. Acad Manage Rev 32:968–985
Testa F, Rizzi F, Daddi T, Marzia Gusmerotti N, Frey M, Iraldo F (2014) EMAS and ISO 14001: the differences in effectively improving environmental performance. J Clean Prod 68:165–173
Thode HC (2002) Testing for normality. Marcel Dekker, New York
Toffel MW (2005) Voluntary environmental management initiatives: smoke signals or smoke screens? PhD dissertation, University of California
Welch EW, Mori Y, Aoyagi-Usui M (2002) Voluntary adoption of ISO 14001 in Japan: mechanism, stages and effects. Bus Strategy Environ 11:43–62

# Adopting ISO Management Standards in Africa: Barriers and Cultural Challenges

**Christian Valery Tayo Tene, Alexander Yuriev and Olivier Boiral**

**Abstract** The adoption of ISO management standards has significantly increased over the last decade in developing countries, especially in Africa. Between 2014 and 2015, the number of certifications in the African continent increased by 20% for ISO 9001 and 19% for ISO 14001. Despite this rapid increase in absolute terms, ISO certifications in Africa remain relatively underdeveloped with about 1% of the total number of ISO 9001 and ISO 14001 certifications worldwide. The objective of this study is to highlight the barriers to the adoption of ISO management standards in the African context and to analyse their appropriateness with the specificities of African cultures and organizational practices. Several barriers are highlighted: the weak institutional framework, the ineffectiveness of donor-funded programs, the lack of human and financial resources, the low participation in the development of ISO management standards, and corruption. In addition, certain specificities of the local cultures in African countries, particularly oral tradition, paternalism, hierarchical distance, collectivism, strong tolerance to uncertainty, and attachment to traditions, may be in opposition to the values embodied in ISO management standards. These specificities require an effort to adapt ISO management systems to African realities in order to promote their substantial rather than symbolic integration within organizations. This work contributes to the existing literature by analyzing the institutional, economic and cultural barriers to the adoption of ISO management standards in Africa. Implications for both professionals and public authorities are also discussed.

C.V. Tayo Tene (✉) · A. Yuriev · O. Boiral
Département de management, Faculté des sciences de l'administration,
Université Laval, Pavillon Palasis-Prince, 2325 rue de la Terrasse,
Local 1638, Quebec, QC G1V 0A6, Canada
e-mail: christian-valery.tayo-tene.1@ulaval.ca

A. Yuriev
e-mail: alexander.yuriev.1@ulaval.ca

O. Boiral
e-mail: Olivier.Boiral@mng.ulaval.ca

 
I. Heras-Saizarbitoria (ed.), *ISO 9001, ISO 14001, and New Management Standards*,
Measuring Operations Performance, https://doi.org/10.1007/978-3-319-65675-5_4

**Keywords** Barriers · ISO 9001 · ISO 14001 · Management standards · ISO certification · African culture

## 1 Introduction

With a total of 1,519,952 certificates issued worldwide in 2015, ISO 9001 and ISO 14001 are, by far, the management standards with the highest number of certifications (ISO 2016a). Many studies have shown that various benefits can result from their adoption, including improved relationships with stakeholders (Poksinska et al. 2003), customer satisfaction (Alemagi et al. 2006; Castka and Corbett 2015), marketing advantages (Poksinska et al. 2003; González-Benito and González-Benito 2005), and improvement in productivity and effectiveness (Jang and Lin 2008; Bansal and Hunter 2003). Among many other reasons, the recognition of these systems in the eyes of stakeholders, and their apparent reputational and practical benefits, have pushed an increasing number of organizations to get ISO 9001 and/or ISO 14001 certifications in the recent years. As a result, the number of certifications has grown exponentially over the last few years and this trend is continuing (ISO 2016a). While the rapid increase at the beginning of the millennium is explained by the popularity of this type of standard in developed countries, developing countries have contributed a great deal to the growth observed in the last decade. Furthermore, annual surveys conducted by ISO indicate that the number of certifications in industrialized countries has stopped growing, or may even have decreased. For example, in Europe, there were 16,000 fewer companies certified with ISO 9001 in 2015 than in 2008 (439,477 and 455,303, respectively). A similar tendency is observed for ISO 14001 in Canada where the number of certified organizations dropped from 1778 in 2012 to 1260 in 2015. This decline in certifications is also observed in South and Central America, as well as in South and Central Asia. Between 2014 and 2015, ISO 14001 certification decreased by 2% in South and Central America, while ISO 9001 certification decreased by 8% in South and Central Asia (ISO 2016a). Conversely, ISO certifications have grown rapidly in many developing countries. For example, in the African continent, ISO 9001 and ISO 14001 certifications increased by 19.8% and 19% respectively between 2014 and 2015 (ISO 2016a). The uneven distribution of ISO certificates across the world raises questions about the applicability of these standards in certain regions (Boiral 2008; d'Iribarne 2003). This chapter discusses how political, social and cultural issues shape the adoption of ISO management standards, and provides a comprehensive overview of various key aspects that hamper companies established in Africa in the implementation of ISO management systems.

The rest of this chapter is organized as follows. First, the dissemination of ISO management standards around the world is discussed from the perspective of standardization and economic globalization. Then, the main barriers to the implementation of ISO management standards observed in Africa are analyzed. Finally, the possible conflicts between the underlying values of the standards and Sub-Saharan cultural traits are explored.

## 2 Standardization and Economic Globalization

Over the years, standardization has become essential, not only to be technically consistent in various sectors, but also for the successful management of organizations. According to Brunsson et al. (2012), standards can be considered to be a dynamic phenomenon, as organizations tend to create their own standards. In the era of globalization, the widespread use of standards became a norm in itself, and there has been a rapid increase in the variety of international voluntary schemes and systems since the 1980s. For instance, certifiable standards that are known worldwide, such as ISO 9001, ISO 14001, SA8000, and Forest Stewardship Council (FSC), do not substitute local laws, but rather tend to support its their implementation (de Oliveira and Pinheiro 2009), and are adopted by companies on a discretionary basis. Most of these standards are certifiable, implying that an audit performed by an accredited organization is mandatory in order to demonstrate compliance with the standards' requirements. The process of auditing, also known as third-party certification, is expected to enhance the credibility of the management system, although its reliability has been questioned in the literature (e.g., Heras-Saizarbitoria et al. 2013; Kouakou et al. 2013). The involvement of an allegedly independent external auditor in the verification process has contributed to the diffusion and popularity of ISO 9001 and ISO 14001 standards (Castka and Corbett 2015).

Many authors conclude that, if properly implemented, ISO management standards are likely to increase economic efficiency, reduce costs and promote international trade (Nadvi and Wältring 2004; Potoski and Prakash 2009). Moreover, they tend to replace local standards that are frequently regarded as barriers to international trade and business development (Clougherty and Grajek 2008; Franceschini et al. 2010; Melnyk et al. 2003). As highlighted by the World Trade Organization (2005, p. 55), "both mandatory and voluntary standards can differ across countries, thus effectively raising a barrier to trade. Such barriers can be removed through harmonization or mutual recognition". Hence, the globalization of the world economy "requires the adoption of international standards that facilitate exchanges and communication between countries" (Boiral 2001, p. 80). According to the mainstream literature, companies implementing the ISO management standards can also improve their effectiveness in terms of administrative processes (Brunsson and Jacobsson 2000), and be more competitive (King et al. 2005), and provide better customer service (Levine and Toffel 2010), or significant social and environmental performance (Tarí et al. 2012).

In principle, standards for quality and environmental management "are intended to be applicable to any organization, regardless of its type or size, or the products

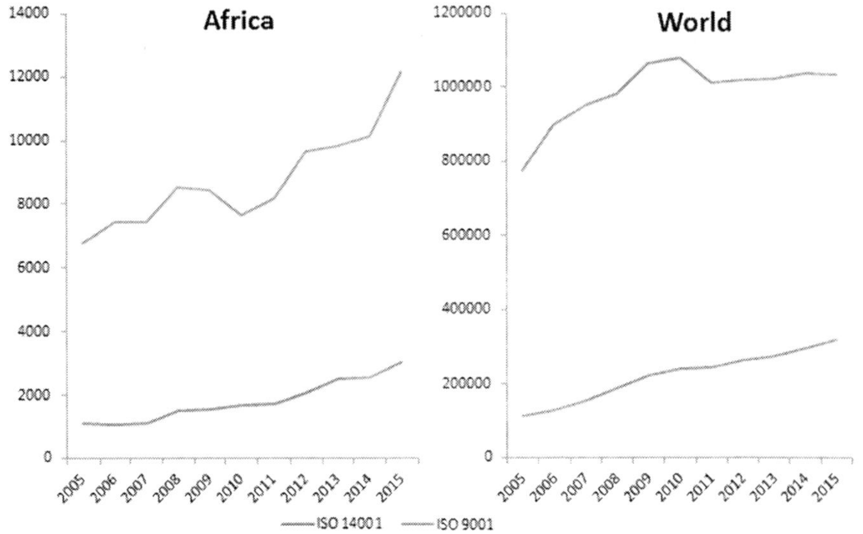


**Fig. 1** Dissemination of ISO 9001 and ISO 14001 in the world and in Africa. *Source* This figure was created based on the data from the ISO survey (2016a)

and services it provides" (ISO 2015a, b, p. 1). From this optimistic perspective, ISO standards can be adopted by firms from both industrialized and developing countries. Such management systems are expected to bring legitimacy and credibility to companies all over the world, and to alleviate commercial barriers. With the growth of the globalized economy and international trade, the number of certified companies worldwide is also increasing. The official statistics (ISO 2016a) confirm this tendency with an increasing number of ISO certificates obtained between 2005 and 2015, particularly in Africa, East Asia, Pacific, Central and South Asia, and the Middle East. For Africa alone, the number of new ISO certificates observed in this period increased by 5000 for ISO 9001 and by 2000 for ISO 14001 (see Fig. 1). While the total number of certificates is obviously increasing, the situation in Africa is not even close to that in other regions. For instance, in 2015, only 1.2% of all ISO 9001 certificates (an increase of only 0.3% since 2005) and 0.9% of all ISO 14001 certificates (a decrease of 0.1% in comparison with 2005) were African.

Moreover, the diffusion of ISO certification in Africa is highly unbalanced, with more than 60% concentrated in two countries: South Africa and Egypt. This situation might be partly explained by the fact that those countries are among the most developed economies in Africa with significant manufacturing and service industries (McKinsey Global Institute 2010). The skewed distribution of ISO management standards is particularly obvious when comparing the 10 African countries with the highest number of certified companies (Table 1). In fact, half of the countries found in this ranking have barely 2% of the total African certifications. South Africa, Egypt, Tunisia and Morocco account for 72% of ISO 9001 and 81% of ISO 14001 certificates found in Africa. Table 1 also reflects the close

**Table 1** Dissemination of ISO management standards across the 10 African countries with the highest number of certified companies

| | Countries | ISO 9001 | % of total in Africa | | Countries | ISO 14001 | % of total in Africa |
|---|---|---|---|---|---|---|---|
| 1 | South Africa | 4346 | 35.76 | 1 | South Africa | 1192 | 39.42 |
| 2 | Egypt | 2484 | 20.44 | 2 | Egypt | 850 | 28.11 |
| 3 | Tunisia | 998 | 8.21 | 3 | Tunisia | 230 | 7.61 |
| 4 | Morocco | 969 | 7.97 | 4 | Morocco | 170 | 5.62 |
| 5 | Kenya | 656 | 5.40 | 5 | Algeria | 102 | 3.37 |
| 6 | Algeria | 569 | 4.68 | 6 | Nigeria | 59 | 1.95 |
| 7 | Mauritius | 239 | 1.97 | 7 | Kenya | 50 | 1.65 |
| 8 | Angola | 208 | 1.71 | 8 | Tanzania | 31 | 1.03 |
| 9 | Nigeria | 202 | 1.66 | 9 | Zimbabwe | 28 | 0.93 |
| 10 | Ivory Coast | 162 | 1.33 | 10 | Ghana | 26 | 0.86 |

*Source* The table was created based on the data from the ISO survey (2016a)

relationship between the adoption of these two management standards: companies certified with ISO 9001—which is usually implemented first—are more inclined to adopt ISO 14001 (King and Lenox 2009), as they are both based on similar documentation and principles (Karapetrovic and Casadesús 2009).

The map of African countries with the highest number of certificates illustrates the discrepancies in terms of ISO certification (see Fig. 2). For example, in 2015, only South Africa and Egypt had more than 2000 ISO 9001 certificates, as well as more than 500 ISO 14001 certificates. Beside these two countries, only four other African states (Tunisia, Morocco, Kenya and Algeria) had more than 500 ISO 9001 certified companies, and only two had between 150 and 500 ISO 14001 certified companies (Morocco and Tunisia). In addition, among the 51 African countries recognized by ISO, 29 have fewer than 30 ISO certificates (e.g., Benin, Burkina Faso, Togo and Rwanda), while others have no ISO certification at all (e.g., Comoros, Sierra Leone and Burundi). It is worth noting that this uneven distribution of ISO management standards is not only found in Africa, as the situation is quite similar in East Asia and South America. For example, among the 165,616 ISO 14001 certificates found in East Asia, 85% are held by two countries: China (114,303) and Japan (26,069). Likewise, 87% of ISO 14001 certifications in South America are held by four countries: Brazil (3113),[1] Colombia (2811), Argentina (1422) and Chile (1214) (ISO 2016a).

The distribution of ISO management standards in Africa also varies considerably depending on the sector concerned. The "Basic Metal and Fabricated Metal products" sector has the highest number of ISO 9001 certifications (see Fig. 3), while the "Food, beverage and tobacco" sector has most ISO 14001 certificates (see

[1]The number in brackets refers to the number of ISO 14001 certificates.

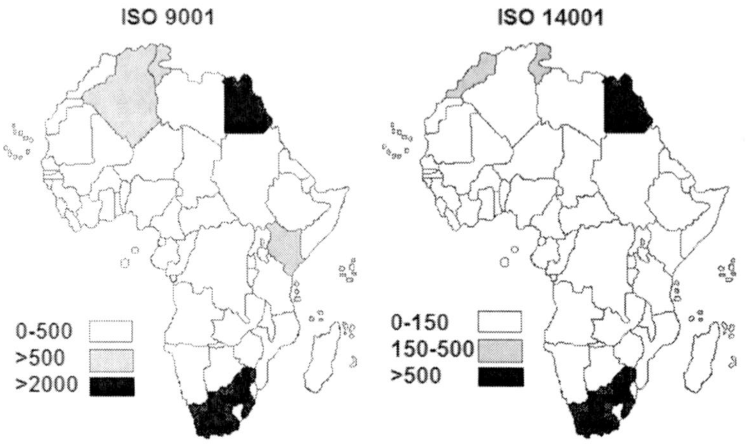


**Fig. 2** African countries with the highest number of certificates. *Source* This map was created based on the data from the ISO survey (2016a)

Fig. 4). Not surprisingly, these two sectors are also among the top 5 industrial sectors with the highest number of ISO 9001 and ISO 14001 certifications worldwide (ISO 2016a). For almost all sectors, the ISO 9001 standard is more widely spread, with the notable exception of the "Mining and quarrying" sector, which has 147 ISO 14001 certifications and only 103 ISO 9001 certifications.

The influence of sector on the adoption of ISO certificates may partly explain the uneven distribution of these standards across African countries. In several cases, more than 70% of ISO certificates for a given industrial sector are found in one or two countries. For example, of a total of 1007 ISO 9001 certifications in the "Basic metal & fabricated metal products" sector, 70% are found in South Africa and 12% in Egypt. The case is the same in the "Chemical products and fibers" sector with 65% of all certificates issued in South Africa and in the "Wholesale & retail trade; repairs of motor vehicles, motorcycles & personal & household goods" sector with 26% of certificates in South Africa, 35% in Kenya and 23% in Egypt. Of all 147 ISO 14001 certificates for the "Mining and quarrying" sector, 73% are issued to companies based in South Africa. On the other hand, several sectors that are considered by McKinsey to have a strong impact on economic growth in Africa have very few ISO certifications. These include tourism (8.7%),[2] financial intermediation (8%), construction (7.5%), transport and telecommunication (7.8%), natural resources (7.1%), and agriculture (5.5%) (McKinsey Global Institute 2010). For example, the "Agriculture, Fishing and Forestry", the "Leather and leather products" and the "Hotels and restaurants" sectors have respectively 90, 17 and 88 ISO 9001 certifications, while the "Machinery and equipment" sector has 346. This

[2]The percentage in brackets refers to the annual growth of the sector according to the McKinsey Global Institute (2010).

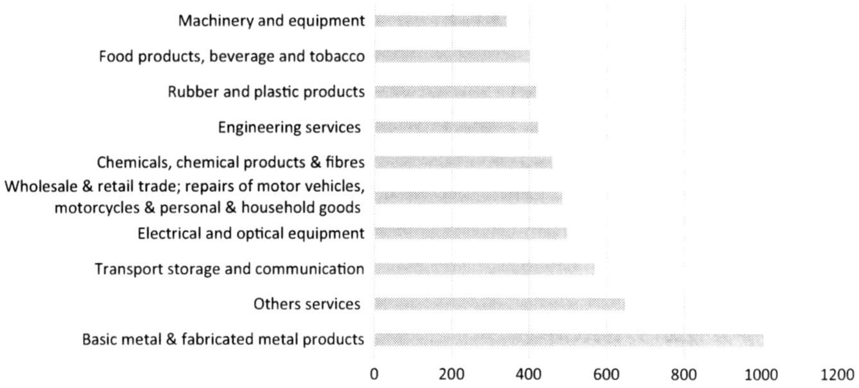


**Fig. 3** ISO 9001 certifications in the main industries in Africa. *Source* This figure was created based on the data from the ISO survey (2016a)

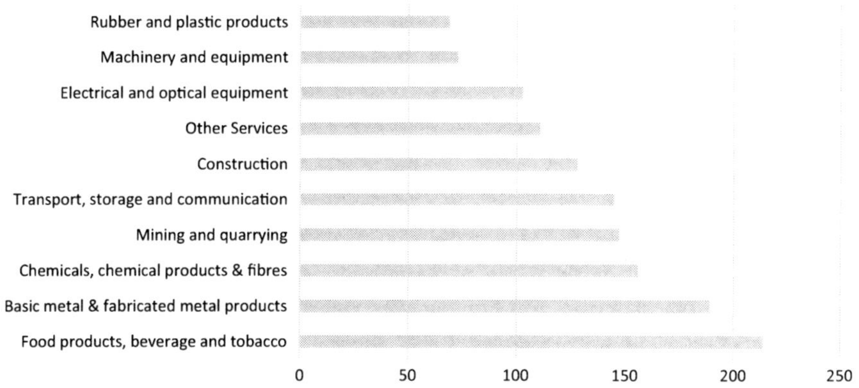


**Fig. 4** ISO 14001 certifications in the main industries in Africa. *Source* This figure was created based on the data from the ISO survey (2016a)

can be partly explained by the predominance of multinationals in some sectors of activity. According to the McKinsey Global Institute (2016), many of the major African-based companies are subsidiaries of multinationals. For example, in Ivory Coast, Senegal and Gabon, almost all ISO-certified companies are subsidiaries of multinationals.

Nevertheless, the use of ISO management standards by those companies can be at odds with local cultural realities. This can lead to some issues for local employees and suppliers who are not familiar with the practices and values incorporated in these standards. The low penetration of ISO management standards in Sub-Saharan Africa may be explained by various economic, institutional and cultural barriers.

# 3 Barriers to ISO Certification in Africa

In this section, potential barriers to the adoption of ISO management standards in Sub-Saharan Africa are explored, especially institutional weakness, limited access to international markets, lack of resources, low participation in the development of standards, inefficiency of programs funded by development partners, corruption, and difficulties related to human resource management.

## *3.1 Institutional Weakness*

Institutional weakness can explain the weak penetration of ISO management standards in the African continent.

First, the involvement of government authorities is essential for the promotion of ISO management standards (Fikru 2014; Massoud et al. 2010). However, the institutional framework for such standards is weak or embryonic in many Sub-Saharan countries, while the situation seems to be different in Egypt and South Africa, which are the continent's leaders for the adoption of ISO management standards. For example, the Egyptian government has put in place an institutional framework with the help of the business community to promote the adoption of ISO 9001 in order to access European and US markets (Magd 2010). Egypt is now ranked among the leading countries in the Arab World for ISO 9001 standards with 2484 certified organizations (ISO 2016a). Despite its rich natural and mineral resources, Africa remains the least industrialized continent (UNECA 2016a). In 2015, Africa's total manufacturing output was valued at around $500 billion and the vast majority of that was concentrated in five countries—Egypt, Morocco, Nigeria, South Africa and Tunisia (McKinsey Global Institute 2016). African countries need more endogenous policies to boost their industrial development (UNECA 2016a; Nzau 2010), which might help to increase the number of certifications (Yumkella and Vinanchiarachi 2003).

Second, national standardization bodies are still nascent in many countries. The role of such organizations is critical: they establish the standardization needs of the states, promote ISO standards, and assure the trustworthiness of certificates (UNIDO 2008). Only 36 countries are members of the African Organization for Standardization (ARSO).[3] A large majority of these bodies are superficial, meaning that they are not operational and, in some cases, exist only "on paper". Consequently, metrology, standardization and conformity assessments are less developed in many countries (UNIDO 2008; Lamy 2012). Generally speaking, African countries need to strengthen or develop reliable standardization bodies to help organizations demonstrate their compliance with ISO management standards. For example, a product may be denied access to a given market, not because it does

[3]http://www.arso-oran.org/members-2/ (Accessed 8 Feb 2017).

not comply with a specific standard, but simply because it is not in a position to demonstrate its compliance (Lamy 2012).

Third, improving compliance with laws and regulations, especially on environmental issues, is one of the main reasons for the promotion of ISO 14001 (Brammer et al. 2012; Alemagi et al. 2006). In many Sub-Saharan countries, regulations related to environmental issues are not enforced, are unclear and, in some cases, are obsolete (Fikru 2014; Massoud et al. 2010; Alemagi et al. 2012). ISO management standards—particularly ISO 14001—cannot be used to improve the compliance with nonexistent or inadequate regulation. Om the other hand, this type of standard can be used as a self-regulation tool to improve the legitimacy of corporate activities and compensate for the lack of strict regulation by supporting management systems that are verified by external auditors (Christmann and Taylor 2006; Heras-Saizarbitoria and Boiral 2013).

Fourth, the lack of incentives impedes the adoption of international standards (Massoud et al. 2010). According to Testa et al. (2016), government incentives (e.g., financial guarantees, tax or fee waivers, and reduction in the frequency of inspections) can be an important driver of ISO certifications. In most African countries, incentives for the adoption of ISO standards are insufficient or unclear, and the bureaucracy of public bodies often obstructs the dissemination of information about the limited incentives available. Many authors highlight the lack of incentives, such as technical and financial support, and tax relief, as a barrier to the adoption of ISO management standards in Africa (e.g., Kehbila et al. 2010; Elbarky and Elzarka 2015).

Fifth, the lack of cooperation between governments and the private sector hampers the dissemination process (Chan 2011). Public administration in some African countries is still very centralized and authoritarian. This situation can make the cooperation between government departments and the private sector more difficult due to administrative and political barriers, and the absence of internal coordination between public agencies. For example, in Cameroon, issues related to the environmental preservation are fragmented and addressed by several public bodies: the Ministry of Industry, the Ministry of the Environment, the Ministry of Energy, the Ministry of Public Health, Services of the Prime Minister, and the Ministry of Culture, as well as by local governments. While additional promotion of ISO standards might seem beneficial, it frequently leads to conflicts of jurisdiction and inertia, making their adoption more complex for interested organizations (Fikru 2014; Alemagi et al. 2012).

Finally, the weakness of civil society in Africa also has a negative influence on the spread of ISO standards. For example, consumer protection groups and environmental movements are still marginal. Given that one of the main drivers in the adoption of ISO management standards is response to institutional pressures (e.g., Agan et al. 2013), the lack of pressure from traditional stakeholders does not encourage the adoption of environmental or quality management standards.

## 3.2 Small Size of Companies and Limited Access to International Markets

One of the problems faced by Sub-Saharan African countries is the weakness of their economic structure, which is characterized by a large number of small and informal enterprises with limited access to international markets (Ndulu et al. 2007; Abor and Quartey 2010; Charmes and Adair 2014). For example, in Ghana and South Africa, SMEs represent more than 90% of all enterprises (Abor and Quartey 2010). Moreover, the informal economy predominates. In African countries, it is estimated that informal economy currently accounts for 40–75% of the GDP and employs between 50 and 80% of the available labor force (Roubaud 2009; Charmes and Adair 2014). These numbers are likely to be underestimates, due to the lack of reliable statistics on this issue (Roubaud 2009). In informal SMEs, labor relations are characterized by casual employment and informal relationships rather than contractual arrangements (Charmes and Adair 2014; Ndulu et al. 2007). These informal enterprises are also characterized by the absence or the lack of formal structure, a focus on the domestic market and limited resources (Mambula 2002; Cook and Nixson 2000). As a result, most organizations are not interested in, or prepared to adopt, international management standards. Most African SMEs are managed by "self-made men" (Kamdem 2002) for whom the ISO certification would not provide significant added value. As a result, even if these entrepreneurs had the resources, it is unlikely that they would adopt ISO management standards, whose usefulness seems, at best, uncertain.

With regard to the limited access to international markets, it is important to note that African countries remain marginal players in world trade with only 2.2% of exports and 4% of imports (CNUCED 2015). Primary products still dominate exports, with fuels accounting for about 40% and agricultural products for more than 25% (Gupta and Yang 2006). African SMEs have difficulty accessing international markets, particularly in Europe and America. This is due to a lack of support from public authorities, and the inability to meet the normative requirements related to products or company management (Gupta and Yang 2006). Since accessing international markets is one of the main reasons for adopting ISO standards (Sakr et al. 2010), difficulties faced by African export firms constitute an obstacle to ISO certification.

## 3.3 Lack of Resources

The lack of resources is one of the main obstacles to the adoption of ISO management standards by African companies (Alemagi et al. 2012; Massoud et al. 2010). According to several authors (e.g., Fikru 2014; Maskus et al. 2005), the costs of quality certification are proportionately substantially higher in developing countries. Although the costs of implementation and certification processes are difficult to measure (Corbett et al. 2005; Henri et al 2014), it ranges from US $12,000 to US$50,000[4] in most African countries. For example, in Cameroon, the cost of an environmental audit is fixed by governmental decree[5] and is estimated at US$10,557, which is very high for most small businesses. Overall, the costs associated with the implementation and certification of ISO management standards is beyond what most African SMEs can afford.

In addition, the lack of expertise and qualifications in ISO management systems has a negative impact on their adoption (UNIDO 2012b; Alemagi et al. 2006; Fikru 2014). This lack of appropriate knowledge is partly related to the low level of literacy and education in many areas of the continent, as there is a positive correlation between education and the adoption of ISO standards (Zeghal and Mhedhbi 2006; Grolleau et al. 2008). Sub-Saharan countries have the highest illiteracy rate in the world,[6] with 38% of adults (about 153 million) being illiterate.[7] Given the focus of ISO certification on documentation and paperwork, this high illiteracy rate is clearly an obstacle to the implementation and monitoring of the standards. In some cases, managers are unable to read and write properly. As a result, companies have to use a translator to educate staff about work methods and procedures.

Moreover, quality and environmental management issues are relatively new to African societies. Education systems in some countries have not yet integrated these issues in their programs. To compensate the lack of local expertise, nationals of countries with people qualified in ISO management standards, such as Cameroon and Ivory Coast, are regularly asked by neighboring states for consulting and auditing services, or hired as quality and environmental managers. The lack of local skills may also partly explain the fact that most certification bodies operating in Africa are foreign companies (e.g., SGS, Bureau Veritas certification, AFAQ certification, DNV, TÜV Rheinland), which contributes to the high costs of ISO certification.

Similarly, the lack of, or poor quality of, infrastructure, particularly electricity, water supply, roads, and information and communication technologies,[8] can represent a serious impediment to the proper implementation and management of ISO management standards. Several African countries experience regular power outages and their sanitation systems remain undeveloped. As a result, some companies cannot implement appropriate measures to improve their environmental performance. For example, wastewater treatment can be a considerable challenge due to the deficiencies of sanitation and waste storage infrastructure. Generally speaking,

---

[4]http://www.iso-certificate-africa.com/ISO-Certification-Cost-in-Africa.html (Accessed 13 March 2017).

[5]Decree No. 2013/0172/PM of 14 February 2013.

[6]http://uis.unesco.org/en/topic/education-africa (Accessed 12 Feb 2017).

[7]http://www.unesco.org/new/fr/dakar/education/literacy/ (Accessed 12 Feb 2017).

[8]https://www.afdb.org/fr/topics-and-sectors/sectors/private-sector/areas-of-focus/infrastructure-finance/ (Accessed 12 Feb 2017).

one of the key factors in Africa's economic growth is "increasing the quantity and quality of the continent's infrastructure" (McKinsey Global Institute 2010).

## 3.4 Low Participation in the Elaboration of ISO Management Standards

African countries are underrepresented in technical committees responsible for the development of ISO standards (Grolleau and Mzoughi 2005; Clapp 2001). Since its creation in 1976, the ISO Technical Committee 176, which is responsible for the quality management standard ISO 9001, has been dominated by experts from Western countries; out of 96 participating countries, only 10 are from Sub-Saharan Africa (Botswana, Ivory Coast, Nigeria, Kenya, Ethiopia, Mali, Rwanda, Tanzania and Zimbabwe).[9] The same situation is observed for ISO 14001.[10] The involvement of African countries, as well as members of civil and professional associations, in technical committee meetings is limited due to the lack of technical expertise. Furthermore, serious financial constraints limit the participation of African experts in international meetings of ISO technical committees. Consequently, ISO management standards are largely based on the values of Western societies. As explained by Clapp (2001, p. 31): "[a] major area of concern regarding the ISO 14000 standards is that of participation in the drafting process. Developing-country governments and environmental NGOs are largely absent from the standard-setting process, while representatives from industrialized countries and Trans-National Corporations dominate the process". In order to restore the balance between developed and developing countries in technical committees, the International Organization for Standardization has recently set a new target for increasing the membership of African, Asian, and South-American experts (ISO 2016b).

## 3.5 Inefficiency of Programs Funded by Development Partners

With the objective of strengthening the capacity of the standardization and management systems of African countries, several programs have been established with financial support from the European Union and technical support from UNIDO. The West Africa Quality Programme (WAQP) was established in 2001

[9]http://www.iso.org/iso/home/standards_development/list_of_iso_technical_committees/iso_technical_committee_participation.htm?commid=53882. (Accessed 9 Feb 2017).

[10]http://www.iso.org/iso/home/standards_development/list_of_iso_technical_committees/iso_technical_committee_participation.htm?commid=54818 (Accessed 9 Feb 2017).

to reinforce regional economic integration and improving trade policies. As part of the Economic Partnership Agreements between Central African countries and the European Union, two major programs were put in place: the "Pilot program for upgrading standardization and quality" (PPUSQ) and the "Quality infrastructure program in central Africa" (CAQIP). The objectives of these programs include the elaboration of regional quality policies, the development of local expertise, the certification of industrial enterprises with ISO 9001, ISO 14001 and ISO 22000 standards, and the accreditation of laboratories in accordance with ISO 17025. Nevertheless, the actual impact of these programs seems relatively low. After many years, there is still a lack of local experts in target countries, and the implementation and auditing of standards is entrusted to foreign professionals. Furthermore, only 7 of the 16 West African countries covered by the WAQP have adopted a policy in relation to standardization and quality (UE 2012). Although these programs have helped to increase the number of ISO certificates by providing funding to companies, their impact on the development of a culture of quality and environmental practices in the public and private sectors seems very low. As a result, many ISO-certified companies decide not to renew their certification after three years. This situation may be partly explained by the poor integration of local realities during the design phase and the management of these programs, as well as the bureaucracy and corruption that make such initiatives more burdensome.

## 3.6 *Corruption*

Although it is not exclusive to Africa, corruption appears to be one of the main barriers to business development, sound governance and substantial implementation of standards in African countries (St-Pierre et al. 2015; UNECA 2016b). Several African countries (e.g., Ivory Coast, Cameroon, Nigeria, Democratic Republic of Congo, Angola, Kenya, Sudan, Tanzania and Zimbabwe) appear regularly in the lower positions in international rankings tracking corruption issues (Transparency International 2016). This acute problem can destroy trust between companies and external stakeholders. It can also extend distrust to ISO certifications, reducing their credibility and signaling value, and thus decreasing the likelihood that firms adopt ISO standards (Montiel et al. 2012). Corruption could favor a superficial integration of ISO management standards due to the ease of obtaining certification. ISO certification may be used for marketing and public relations purposes only, while internal practices remain untouched (Heras-Saizarbitoria and Boiral 2015; Iatridis and Kesidou 2016).

Moreover, corruption at the level of enterprises and governments alike can also affect the effectiveness of public policies implemented to promote the adoption of ISO standards (Alemagi et al. 2012). Similarly, collusion between consultants and auditors can encourage the development of "fake certificates" (Fryxell et al. 2004; Heras-Saizarbitoria et al. 2013) delivered by unscrupulous auditors or non-accredited consultants. In addition, it is common to observe business relationships between consultants in charge of the implementation of ISO standards and certification bodies in charge of the certification process. The problem of corruption also leads to a lack of transparency of certification bodies (UNIDO 2012a), affecting the whole standardization system in Africa.

## *3.7 Difficulties Related to Human Resource Management*

Human resource management practices in African companies are rarely in line with the requirements of international standards. For example, although ISO 9001 requires that the company must "ensure that people have the right education, training or experience" (ISO 2015a, p. 11), recruitment is often informal and based on co-optation with an emphasis on ethnicity, family or religious affiliation rather than actual competence (Chalus-Sauvannet and Noguera 2010; Nkakleu 2016; Apitsa and Amine 2014). Similarly, although ISO 14001 requires companies to "determine training needs associated with its environmental aspects and its environmental management system" (ISO 2015b, p. 11), staff training is rarely the priority of top managers, especially in SMEs that have limited financial resources (St-Pierre et al. 2015; Ishengoma and Kappel 2008). In cases where training is offered, it is often unsuited to the workplace realities due to the lack of rigor in the identification of training needs and the lack of follow-up (Chalus-Sauvannet and Noguera 2010). In addition, remuneration and job classification are often fixed in a discretionary manner by the manager, irrespective of legal requirements (Nkakleu 2016; Apitsa and Amine 2014). It is quite common that qualified people recruited to manage quality and/or environmental issues are underpaid considering their responsibilities (St-Pierre et al. 2015; Ishengoma and Kappel 2008). As explained by the quality manager of a multinational firm based in Cameroon interviewed by one of the authors, "This reality reflects the little interest of the top management for the function and activities related to quality and the environment. This results in turnover of employees which hinders the continuity of the management system in place". Moreover, human resource management processes need to be more formalized to facilitate the monitoring and verification of evidence in internal and external audits. In African SMEs, human resource management frequently remains informal and poorly documented (Chalus-Sauvannet and Noguera 2010; Nkakleu 2016; Apitsa and Amine 2014).

The barriers mentioned above are, to some extent, related to cultural aspects, which raises the question of the compatibility between ISO management standards and traditional African values.

## 4 Are ISO Management Standards Adapted to African Cultures?

According to Hofstede (1993), culture operates as a sort of mental software that shapes the psychological structures from which behaviors flow. Culture comes from the sedimentation of national traditions over a long period of time (d'Iribarne 2003). Adaptation to local cultural specificities is an important aspect for the success of organizational projects and management practices. Some cultural traits of Africans are considered to be major obstacles to business development and the adoption of new practices on the continent (Kamdem 2002; Verna and Finifter 2011). The same is true of the adoption of certifiable standards such as ISO 9001 and ISO 14001 (Boiral 2008). Although the literature on African management is fairly scattered (Walsh 2015), various authors have described its key characteristics (e.g., Boiral and Mboungou 2004; Kamdem 2002; Zadi Kessy 1998; Mutabazi 2006; Beugré and Offodile 2001). These traits include: power distance, paternalism, collectivism, risk tolerance, polychronic time, interpersonal communication, palaver, magico-religious practices and attachment to traditions. Table 2 compares common peculiarities of local culture in many African countries with the values conveyed by ISO management standards.

As illustrated in Table 2, the peculiarities of African management can undermine the adoption of international ISO standards. For example, the latest version of ISO 14001, published in 2015, states that: "Top management shall ensure that the responsibilities and authorities for relevant roles are assigned and communicated within the organization" (ISO 2015b, p. 8). It would seem to be difficult to reconcile the centralization of management frequently observed in the African region with this requirement. Power distance might alter internal communication, jeopardize the involvement of staff in company projects and, in some cases, even favor the centralization of powers. In many SMEs, managers play multiple roles and can simultaneously be in charge of purchasing, finance and human resources. The focus on specific roles in ISO management systems raises questions about this multitasking rationale. As the power is expected to be decentralized across multiple individuals in the company, adopting management standards may appear to go against the traditional leader's authority. This remark also applies to the verification process conducted by external auditors, which questions the centralized power exerted by traditional leaders. Moreover, the paternalistic logic, which is based on trust, traditions and the protective role of managers, does not promote the same values as ISO management standards which rely on formalized relationships in the organization based on procedures, monitoring of objectives and clear allocation of roles and responsibilities (see Table 2). Similarly, collectivism, which emphasizes mutual assistance, solidarity, and belonging to the family and ethnic group, promotes informal and tacit interpersonal relationships rather than the formal role definition proposed by ISO management standards. The elastic and polychronic

**Table 2** Comparison of the traits of African management and the values conveyed by ISO management standards

| Cultural traits | Main features of African management | Values conveyed by the ISO management standards |
|---|---|---|
| Power distance | High level of power distance (House et al. 2004; Hofstede 1993); predominance of domination-subordination relationships (Kamdem 2002; Henry 2008) and authoritarian leadership (Verna and Finifter 2011); hierarchical relationship between people with power and employees (Aycan 2005; Boiral 2008); highly centralized power structures (d'Iribarne 2003; Zadi Kessy 1998; Hernandez and Kamdem 2007); autocratic leadership style (Beugré and Offodile 2001) | The power relations in the ISO management standards are based on a formalized management system: definition of policies, procedures, objectives, etc. Through this formalization, they call into question the personal authority of the leader |
| Paternalism | Superiors provide care and protection to their subordinates (Pellegrini et al. 2010); highly emotional link between leaders and subordinates (Etcheu 2013; Boiral 2008) | Impersonal relationships based on respect for procedures, role-definition, clarification of responsibilities and authority over management systems |
| Collectivism | Solidarity, loyalty, a strong sense of belonging (Boiral 2008); promotion of interpersonal relations (Levesque et al. 2004); mutual assistance within ethnic groups (Kamdem 2002; Mutabazi 2006; d'Iribarne 2003); humanism and community involvement (West 2014; Seny Kan et al. 2015); common references that simplify conversations (Nkakleu 2009; Mutabazi 2006); the sharing of tacit knowledge (Boiral and Mboungou 2004) | The ISO standards require the definition of roles and responsibilities according to the management system in place rather than the social standing of individuals. Since the ISO systems are formalized, they involve explicit communication with abundance of written documents. Tacit and informal aspects tend to be overlooked |
| Tolerance in the face of uncertainty | High tolerance to uncertainty and ambiguity (Kamoche et al. 2012; Hernandez and Kamdem 2007; d'Iribarne 2003); low propensity to avoid uncertainty through rigid practices (Boiral 2008; Verna and Finifter 2011; Mutabazi 2006) | The formalism of ISO standards reflects a lack of tolerance for uncertainty. The gap between the requirements of these standards and their achievement are considered to be non-compliance |

(continued)

**Table 2** (continued)

| Cultural traits | Main features of African management | Values conveyed by the ISO management standards |
|---|---|---|
| Polychronous time | Time is not gained or wasted, it is shared with others (Kamdem 2010; Verna and Finifter 2011); time has no relation to money (Zadi Kessy 1998); as a result, frequency of delays and less rigorous planning (Zadi Kessy 1998; Kamdem 2010) | The ISO systems emphasize the administrative and measurable dimensions of time by obscuring its social dimension. Business planning and monitoring of the goals are generally defined through a linear vision of time |
| Role of the palaver | The palaver is the "jurisdiction of speech" (Bidima 2009), which is referred to as a form of democracy in Africa; it helps to solve interpersonal problems through the right use of words (Beugré and Offodile 2001; Kamdem 2002); it exists to rebuild relationships with others (d'Iribarne 2003) | The palaver culture seems difficult to integrate in ISO management standards. The privileged mode of communication of ISO standards is formal documents, which encourages the monitoring of the system in place. This requirement of formalism is not aligned with the oral tradition of African cultures |
| Magico-religious practices | Use of magical rituals and witchcraft to influence labor relations (Kamdem 2002; Boiral and Mboungou 2004); religious/witchcraft practices are used to strengthen one's power (d'Iribarne 2003; Henry 2008), as well as to combat fraud (Fotoke 2007) | Magico-religious practices appear to be in opposition to the formalism and rationality of ISO management standards |
| Attachment to traditions | Traditions help to find the meaning of life, the essence of which is deeply rooted in local culture (Bidima 2009; Kamdem 2002); respect for elders (Worou 2010; Bidima 2009; Henry 2008); correlation between age and wisdom (Beugré and Offodile 2001), resistance to management practices that do not take local traditions into account (Kamdem 2002; d'Iribarne 2003) | ISO management standards offer little flexibility in relation to local traditions, although the last version of ISO 14001 emphasizes the identification and adaptation to the context of organizations. Overall, ISO management standards represent an international reference framework that does not take into account the adaptation to local traditions |

Adapted from Boiral and Mboungou (2004, p. 5)

attitude towards time in Africa (e.g., Kamdem 2002; Zadi Kessy 1998; Beugré and Offodile 2001) may partly explain a lack of respect for deadlines, frequent delays in the execution of tasks and an inability to plan future processes precisely. These pitfalls can undermine the efficient implementation of ISO management standards, which emphasise the critical role of planning and continuous improvement over time.

Furthermore, ISO 14001 and ISO 9001 tend to encourage a cautious and preventive attitude: "The organization can ensure this by determining its risks and opportunities that need to be addressed and planning action to address them. These risks and opportunities can be related to environmental aspects, compliance obligations, other issues or other needs and expectations of interested parties" (ISO 2015b, p. 22). This trend reflects the low risk tolerance that characterizes Western society (see Table 2). In contrast with this, African companies are characterized by a higher tolerance of uncertainty (Turki 2009; Boiral 2008). The implementation of risk management practices would require a profound cultural change which seems unlikely given the deeply rooted traditional and religious beliefs, such as "God controls everything" or "We must accept God's will". This type of belief can partly explain why the principle of prevention, which is supposed to be crucial for the adoption of ISO management standards, is practically nonexistent in most African companies.

Although ISO management standards encourage communication and information sharing within organizations, the written form is recommended in order to facilitate follow-ups and the certification process. In many Sub-Saharan organizations, communication, information exchange and definition of tasks are usually done through oral conversations rather than documented procedures. Also, ISO management standards emphasize the need to collect written versions of various rules and to keep a record of documents. The predominance of informal relationships in African countries makes such codification more difficult to implement and to verify through auditing procedures (Kouakou et al. 2013).

Despite these cultural and institutional barriers, some researchers consider it possible to implement ISO management standards efficiently in African companies (Zadi Kessy 1998; d'Iribarne 2003). Africans tend to respect established rules, which could facilitate the adoption of international standards and mitigate some of the perverse effects related to local cultures. Overall, when adopted properly and with respect to African specificities, ISO management systems can represent an opportunity to promote a self-regulation rationale and to offset the weakness of regulatory and institutional frameworks, while enabling companies to implement more efficient environmental and quality practices.

## 5 Conclusion

The objectives of this chapter were to describe barriers to the adoption of ISO management systems in Sub-Saharan countries and to analyze the appropriateness of these standards in the African context. Although the number of ISO certified companies is growing worldwide, African organizations seem reluctant or unable to adopt this type of standard. The review of the barriers associated with the adoption of ISO standards in the African context helps to clarify key challenges with the implementation of international standards. These barriers include various aspects, such as institutional weakness, unclear industrial policies, lack of governmental incentives, poor dissemination of information and knowledge about the standards, quasi-absence of reliable local certification bodies, and the predominance of informal SMEs shaped by traditional culture. Lack of human and financial resources also represents a significant barrier to the adoption of ISO standards. African economies are dominated by SMEs and family businesses that cannot afford the costs of the implementation and external verification of ISO management standards. Collusion between state officials, inspectors and managers, as well as various dubious practices of local auditors, are a fertile breeding ground for the development of fake certificates and tend to cast a shadow on the external verification industry in Africa. This negative perception of third-party verification creates additional difficulties for African companies wishing to access foreign markets. Although the attractiveness of international markets for certain organizations can be a sufficient motivation in itself to get certified, it rarely translates into practices, for the reasons explored in this chapter. Apart from these acute issues, consumer and environmental protection movements are still embryonic in several African countries. Actually, they lack the human and financial resources, as well as the support from the authorities, to put pressure on enterprises to adopt new practices. Overall, the lack of institutional pressure for the adoption of ISO management standards tends to encourage "business as usual".

Moreover, while ISO management standards are shaped by Western values (see Table 2), the local culture of many African countries is dramatically different and appears to be in opposition to some ISO requirements. Collectivism, paternalism, high risk tolerance, the importance of palaver and attachment to traditions are characteristic features of African management. These cultural aspects contrast with the dominant Western values characterized by individualism, contractual formalism and low tolerance of risk (d'Iribarne 2003; Hernandez and Kamdem 2007; Henry 2008). While ISO management standards tend to convey Western management values, based on formalized and systematic management practices, most African organizations remain essentially based on informal and mostly oral means of communication.

The misalignment of ISO standards with African traditions is not necessarily an insurmountable barrier. The flexibility of ISO requirements can help in adapting the standard to the local culture (Beugré and Offodile 2001). Customized training and personalized implementation can also help this adaptation process (Zadi Kessy 1998; d'Iribarne 2003). Nevertheless, the cultural barriers may encourage a superficial and symbolic rather than substantial integration of ISO standards in many African organizations. On the one hand, this lack of internalization of ISO management standards is fairly common in Western companies (Christmann and Taylor 2006) and, one can assume that, due to the cultural differences and the other barriers mentioned, the situation may be even worse in most African regions. On the other hand, the symbolic integration of ISO management standards can be seen as a way of alleviating cultural barriers to enable African organizations to embrace the international trend toward increasing standardization. In any case, the adoption of ISO standards could facilitate the access of African organizations to international markets and stimulate certain sectors of activity.

By highlighting the various barriers to the adoption of ISO standards and the cultural differences between African values and those conveyed by these standards, this chapter provides useful information for ISO practitioners and managers. It could help them to anticipate certain difficulties in the implementation of these standards and to develop policies and public programs for the promotion of ISO standards that are more suitable for African realities. Exploring the most efficient way of establishing common ground between the formalized systems offered by ISO management standards and African cultural specificities offers many opportunities for future research. For example, an interesting avenue of research would be to analyze how the local culture shapes the integration of ISO management standards in African organizations coming from different regions. Given the scarcity of research on the internalization of these standards in Africa, the study of cultural factors influencing the superficial or substantial integration of ISO standards is also an interesting avenue of research. In addition, the management of relationships with interested parties, which is an important aspect of the 2015 versions of ISO 9001 and ISO 14001, could be analyzed through cases studies. Corruption, which is endemic to several Sub-Saharan African countries, and its impact on the development of "fake certificates", could also be investigated. The practices needed for the adaptation of international certification bodies to African cultures, as well as the context of corruption, could also be investigated. Finally, institutional and cultural barriers to the adoption of the ISO management standards should be further analyzed through empirical studies involving various stakeholders located in different African regions.

## References

Abor J, Quartey P (2010) Issues in SME development in Ghana and South Africa. Int Res J Finance Econ 39(6):215–228

Agan Y, Acar MF, Borodin A (2013) Drivers of environmental processes and their impact on performance: a study of Turkish SMEs. J Clean Prod 51:23–33

Alemagi D, Oben PM, Ertel J (2006) Implementing environmental management systems in industries along the Atlantic coast of Cameroon: drivers, benefits and barriers. Corp Soc Responsib Environ Manage 13(4):221–232

Alemagi D, Hajjar R, David S, Kozak RA (2012) Benefits and barriers to certification of community-based forest operations in Cameroon: an exploratory assessment. Small Scale For 11(4):417–433

Apitsa SM, Amine A (2014) L'ethnicité, un levier d'action pertinent du management interculturel des ressources humaines? Manage Avenir 4:13–31

Aycan Z (2005) The interplay between cultural and institutional/structural contingencies in human resource management practices. Int J Hum Resour Manage 16(7):1083–1119

Bansal P, Hunter T (2003) Strategic explanations for the early adoption of ISO 14001. J Bus Ethics 46:289–299

Beugré CD, Offodile OF (2001) Managing for organizational effectiveness in Sub-Saharan Africa: a culture-fit model. Int J Hum Resour Manage 12(4):535–550

Bidima JG (2009) Philosophie et traditions dans l'espace public africain. Cah Sens Public 2:113–132
Boiral O (2001) ISO 14001 certification in multinational firms: the paradoxes of integration. Glob Focus 13(1):79–94
Boiral O (2008) Les pays du sud à l'épreuve des normes ISO: vers un sous-développement durable? Manage Int 12(2):49
Boiral O, Mboungou G (2004) Gérer dans la tourmente: le management des entreprises formelles au Congo-Brazzaville. Manage Int 8(2):33
Brammer S, Hoejmose S, Marchant K (2012) Environmental management in SMEs in the UK: practices, pressures and perceived benefits. Bus Strategy Environ 21(7):423–434
Brunsson N, Jacobsson B (2000) A world of standards. Oxford University Press, Oxford
Brunsson N, Rasche A, Seidl D (2012) The dynamics of standardization: three perspectives on standards in organization studies. Organ Stud 33(5–6):613–632
Castka P, Corbett CJ (2015) Management systems standards: diffusion, impact and governance of ISO 9000, ISO 14000, and other management standards. Found Trends Technol Inf Oper Manage 7(3–4):161–379
Chalus-Sauvannet MC, Noguera F (2010) Étude empirique au sein d'une entreprise subsaharienne. Rev Fr Gest 5:15–31
Chan ES (2011) Implementing environmental management systems in small-and medium-sized hotels: obstacles. J Hosp Tour Res 35(1):3–23
Charmes J, Adair P (2014) L'inconstant caméléon, ou comment appréhender l'informel? Mondes Dév 2:7–16
Christmann P, Taylor G (2006) Firm self-regulation through international certifiable standards: determinants of symbolic versus substantive implementation. J Int Bus Stud 37(6):863–878
Clapp J (2001) ISO environmental standards: industry's gift to a polluted globe or the developed world's competition-killing strategy? Yearb Int Co Oper Environ Dev 2002:27–33
Clougherty JA, Grajek M (2008) The impact of ISO 9000 diffusion on trade and FDI: a new institutional analysis. J Int Bus Stud 39(4):613–633
CNUCED (Conférence des nations unies sur le commerce et le développement) (2015) Le développement économique en Afrique Rapport 2015. Nations-Unies, New York. Retrieved from http://unctad.org/fr/PublicationsLibrary/aldcafrica2015_fr.pdf
Cook P, Nixson F (2000) Finance and small and medium-sized enterprise development. Institute for Development Policy and Management, University of Manchester, Manchester
Corbett CJ, Montes-Sancho MJ, Kirsch DA (2005) The financial impact of ISO 9000 certification in the United States: an empirical analysis. Manage Sci 51(7):1046–1059
d'Iribarne P (2003) Le tiers-monde qui réussit: nouveaux modèles. Éditions Odile Jacob, Paris
de Oliveira OJ, Pinheiro CRMS (2009) Best practices for the implantation of ISO 14001 norms: a study of change management in two industrial companies in the Midwest region of the state of São Paulo-Brazil. J Clean Prod 17(9):883–885
Elbarky S, Elzarka S (2015) A green supply chain management migration model based on challenges faced in Egypt. In: International conference on industrial engineering and operations management (IEOM). IEEE, pp 1–8
Etcheu J (2013) L'implication du paternalisme dans la sélection du personnel à la formation: une étude de cas des PME camerounaises. Humanisme Entrep 2:73–88
Fikru MG (2014) International certification in developing countries: the role of internal and external institutional pressure. J Environ Manage 144:286–296
Fotoke M (2007) Le recours à la religion pour lutter contre les fraudes. Le cas d'une entreprise beninoise. In: Nizet J, Pichault F (eds) Les performances des organisations africaines: pratiques de gestion en contexte incertain, Éditions l'Harmattan, pp 107–120
Franceschini F, Galetto M, Maisano D, Mastrogiacomo L (2010) Clustering of European countries based on ISO 9000 certification diffusion. Int J Qual Reliab Manage 27(5):558–575
Fryxell GE, Chung SS, Lo CW (2004) Does the selection of ISO 14001 registrars matter? Registrar reputation and environmental policy statements in China. J Environ Manage 71 (1):45–57

González-Benito J, González-Benito O (2005) An analysis of the relationship between environmental motivations and ISO 14001 certification. Br J Manage 16:133–148

Grolleau G, Mzoughi N (2005) L'élaboration des normes: un «nouvel» espace de compétition? Une application à la norme ISO 14001. Rev Écon Ind 111(1):29–56

Grolleau G, Lamri J, Mzoughi N (2008) Déterminants de la diffusion internationale de la norme ISO 14001. Econ Prévis 4:123–138

Gupta S, Yang Y (2006) Les freins au commerce en Afrique. Finance Dév. Retrieved from https://www.imf.org/external/pubs/ft/fandd/fre/2006/12/pdf/Gupta.pdf

Henri JF, Boiral O, Roy MJ (2014) The tracking of environmental costs: motivations and impacts. Eur Account Rev 23(4):647–669

Henry A (2008) Culture et gestion au Cameroun: le respect des rituels, source d'une entente amicale. In: Gestion en Contexte Interculturel: Approches, Problématiques, Pratiques et Plongées. Presses de l'Université Laval et Télé-université, Québec

Heras-Saizarbitoria I, Boiral O (2013) ISO 9001 and ISO 14001: towards a research agenda on management system standards. Int J Manage Rev 15(1):47–65

Heras-Saizarbitoria I, Boiral O (2015) Symbolic adoption of ISO 9000 in small and medium-sized enterprises: the role of internal contingencies. Int Small Bus J 33(3):299–320

Heras-Saizarbitoria I, Dogui K, Boiral O (2013) Shedding light on ISO 14001 certification audits. J Clean Prod 51:88–98

Hernandez ÉM, Kamdem E (2007) Universalité ou contingence de l'enseignement de la gestion Le cas de l'Afrique. Rev Fr Gest 9:25–41

Hofstede G (1993) Cultural constraints in management theories. Acad Manage Exec 7(1):81–94

House RJ, Hanges PJ, Javidan M, Dorfman PW, Gupta V (eds) (2004) Culture, leadership, and organizations. The GLOBE Study of 62 Societies, Thousand Oaks

Iatridis K, Kesidou E (2016) What drives substantive versus symbolic implementation of ISO 14001 in a time of economic crisis? Insights from Greek manufacturing companies. J Bus Ethics 1–19

Ishengoma EK, Kappel R (2008) Business constraints and growth potential of micro and small manufacturing enterprises in Uganda. GIGA J Res Prog 5(78):1–29

ISO (International Organization for Standardization) (2015a) ISO 9001:2015 quality management systems—requirements with guidance for use. ISO (International Organization for Standardization), Geneva

ISO (International Organization for Standardization) (2015b) ISO 14001:2015 environmental management systems—requirements with guidance for use. ISO (International Organization for Standardization), Geneva

ISO (International Organization for Standardization) (2016a) The ISO survey of management system standard certifications 2015. ISO Central Secretariat, Geneva

ISO (International Organization for Standardization) (2016b) Plan d'action de l'ISO pour les pays en développement 2016–2020. Secrétariat central de l'ISO, Geneve. Retrieved from http://www.iso.org/iso/fr/iso_action_plan_2016-2020_en_ld.pdf

Jang W, Lin C (2008) An integrated framework for ISO 9000 motivation, depth of ISO implementation and firm performance. The case of Taiwan. J Manuf Technol Manage 19:194–216

Kamdem E (2002) Management et interculturalité en Afrique. Expérience camerounaise. Les Presses de l'Université Laval/L'Harmattan, Saint-Nicolas/Paris, coll. «Sciences de l'administration»

Kamdem E (2010) Itinéraire de recherche qualitative sur les temporalités en Afrique. Recherches Qualitatives, Hors serie no. 8, pp 61–75

Kamoche K, Chizema A, Mellahi K, Newenham-Kahindi A (2012) New directions in the management of human resources in Africa. Int J Hum Resour Manage 23(14):2825–2834

Karapetrovic S, Casadesús M (2009) Implementing environmental with other standardized management systems: scope, sequence, time and integration. J Clean Prod 17:533–540

Kehbila AG, Ertel J, Brent AC (2010) Corporate sustainability, ecological modernization and the policy process in the South African automotive industry. Bus Strategy Environ 19(7):453–465

King A, Lenox MJ (2009) Lean and green? An empirical examination of the relationship between lean production and environmental performance. Prod Oper Manage 10:244–256
King AA, Lenox MJ, Terlaak A (2005) The strategic use of decentralized institutions: exploring certification with the ISO 14001 management standard. Acad Manage J 48(6):1091–1106
Kouakou D, Boiral O, Gendron Y (2013) ISO auditing and the construction of trust in auditor independence. Account Audit Account J 26(8):1279–1305
Lamy P (2012) Avant-propos. L'OMC et les normes internationales. In: Annales des Mines-Responsabilité et environnement, no. 3. ESKA, pp 7–11
Levesque M, Blais MR, Hess U (2004) Motivation, comportements organisationnels discrétionnaires et bien-être en milieu Africain: quand le devoir oblige? Can J Behav Sci 36 (4):321
Levine DI, Toffel MW (2010) Quality management and job quality: how the ISO 9001 standard for quality management systems affects employees and employers. Manage Sci 56(6):978–996
Magd H (2010) Quality management standards (QMS) implementation in Egypt: ISO 9000 perspectives. Glob Bus Manage Res Int J 2(1):57–68
Mambula C (2002) Perceptions of SME growth constraints in Nigeria. J Small Bus Manage 40 (1):58
Maskus K, Otsuki T, Wilson J (2005) The costs of complying with foreign product standards for firms in developing countries. World Bank policy research working paper no. 3590
Massoud MA, Fayad R, El-Fadel M, Kamleh R (2010) Drivers, barriers and incentives to implementing environmental management systems in the food industry: a case of Lebanon. J Clean Prod 18(3):200–209
McKinsey Global Institute (2010) Lions on the move: the progress and potential of African economies. Retrieved from file:///C:/Users/Christian/Downloads/MGI_Lions_on_the_move_african_economies_full_report%20(1).pdf
McKinsey Global Institute (2016) Lions on the move II: realizing the potential of Africa's economies. Retrieved from file:///C:/Users/Christian/Downloads/MGI-Lions-on-the-Move-2-Full-report-September-2016v2%20(1).pdf
Melnyk SA, Sroufe RP, Calantone R (2003) Assessing the impact of environmental management systems on corporate and environmental performance. J Oper Manage 21(3):329–351
Montiel I, Husted BW, Christmann P (2012) Using private management standard certification to reduce information asymmetries in corrupt environments. Strateg Manage J 33(9):1103–1113
Mutabazi E (2006) Face à la diversité des cultures et des modes de gestion: le modèle circulatoire de management en Afrique. Manage Avenir 4:179–197
Nadvi K, Wältring F (2004) Making sense of global standards. In: Schmitz H (ed) Local enterprises in the global economy. Edward Elgar, Cheltenham, pp 53–94
Ndulu B, Chakraborti L, Lijane L, Ramachandran V, Wolgin J (2007) Challenges of African growth: opportunities, constraints and strategic directions. World Bank, Washington
Nkakleu R (2009) Quand la tontine d'entreprise crée le capital social intra-organisationnel en Afrique: Une étude de cas. Manage Avenir 7:119–134
Nkakleu R (2016) Les pratiques de GRH des PME africaines sont-elles toujours informelles? Une analyse contextualiste. Quest Manage 1:83–104
Nzau M (2010) Africa's industrialization debate: a critical analysis. J Lang Technol Entrep Africa 2(1):146–165
Pellegrini EK, Scandura TA, Jayaraman V (2010) Cross-cultural generalizability of paternalistic leadership: an expansion of leader-member exchange theory. Group Org Manage 35(4):391–420
Poksinska B, Dahlgaard JJ, Eklund JAE (2003) Implementing ISO 14000 in Sweden: motives, benefits and comparisons with ISO 9000. Int J Qual Reliab Manage 20:585–606
Potoski M, Prakash A (2009) Information asymmetries as trade barriers: ISO 9000 increases international commerce. J Policy Anal Manage 28(2):221–238
Roubaud F (2009) La production statistique sur le secteur informel en Afrique: quels enseignements et quelles perspectives? Afr Stat J 9:399–421

Sakr DA, Sherif A, El-Haggar SM (2010) Environmental management systems' awareness: an investigation of top 50 contractors in Egypt. J Clean Prod 18(3):210–218
Seny Kan KA, Apitsa SM, Adegbite E (2015) "African management": concept, content and usability. Soc Bus Rev 10(3):258–279
St-Pierre J, Foleu L, Abdulnour G, Nomo S, Fouda M (2015) SME development challenges in cameroon: an entrepreneurial ecosystem perspective. Transnatl Corp Rev 7(4):441–462
Tarí JJ, Molina-Azorín JF, Heras I (2012) Benefits of the ISO 9001 and ISO 14001 standards: a literature review. J Ind Eng Manage 5(2):297
Testa F, Heras-Saizarbitoria I, Daddi T, Boiral O, Iraldo F (2016) Public regulatory relief and the adoption of environmental management systems: a European survey. J Environ Planning Manage 59(12):2231–2250
Transparency International (2016) Corruption perceptions index 2016. Retrieved from http://www.transparency.org/whatwedo/publication/corruption_perceptions_index_2016
Turki A (2009) Les comportements écologiques des dirigeants des entreprises tunisiennes. VertigO-la revue électronique en sciences de l'environnement 9(2):1–9
UE (Union Européenne) (2012) Les résultats clés du Programme Qualité de l'Afrique de l'Ouest (PQAO). Retrieved from http://eeas.europa.eu/archives/delegations/burkina_faso/documents/press_corner/20130827_waqp_fr.pdf
UNECA (United Nation Economic Commission for Africa) (2016a) Greening Africa's industrialization. Retrieved from http://www.un.org/en/africa/osaa/pdf/pubs/2016era-uneca.pdf
UNECA (United Nation Economic Commission for Africa) (2016b) Mesurer la corruption en Afrique: prendre en compte la dimension internationale. Retrieved from http://www.un.org/africarenewal/sites/www.un.org.africarenewal/files/Rapport_sur_la_gouvernance_en_Afrique%20IV.pdf
UNIDO (2008) Fast forward: national standards bodies in developing countries. Retrieved from http://www.unido.org/fileadmin/user_media/Publications/documents/fast_forward.pdf
UNIDO (2012a) ISO 9001—its relevance and impact in Asian developing economies. Retrieved from http://www.unido.org/fileadmin/user_media/Publications/Pub_free/ISO%209001%20Impact%20Survey-eBook_ver2.pdf
UNIDO (2012b) Nouvelle politique industrielle de la République de Côte d'Ivoire Phase I: diagnostic du secteur industriel et du cadre institutionnel. Retrieved from https://www.unido.org/fileadmin/user_media/Research_and_Statistics/Policy_Advisory_Services/Industrial_Policy_Report_Cote_d_Ivoire_Oct_2012.pdf
Verna G, Finifter R (2011) L'entreprise africaine: essai sur la mal gouvernance (Nouvelle collection Nord-Sud). Presses de l'Universite Laval, Québec
Walsh JP (2015) Organization and management scholarship in and for Africa... and the World. Acad Manag Perspect 29(1):1–6
World Trade Organization (2005) World trade report: exploring the links between trade, standards and the WTO. Retrieved from https://www.wto.org/english/res_e/booksp_e/anrep_e/world_trade_report05_e.pdf
Worou D (2010) Impact des pratiques de gestion des ressources humaines sur l'acceptation de l'ERP dans les entreprises en Afrique: cas de deux entreprises en Afrique de l'Ouest. In: Tidjani B, Kamdem E (eds) Gérer les ressources humaines en Afrique: entre processus sociaux et pratiques organisationnelles. Éditions Management et Société, Caen, pp 121–144
Yumkella K, Vinanchiarachi J (2003) Leading issues on Africa's path to industrialization: the role of support systems and instruments. J Afr Econ 12(1):30–40
Zadi Kessy M (1998) Culture africaine et gestion de l'entreprise moderne. CEDA, Abidjan
Zeghal D, Mhedhbi K (2006) An analysis of the factors affecting the adoption of international accounting standards by developing countries. Int J Account 41(4):373–386

# Implementing Service Management Standards: Motivations and Key Factors

**Cots Santi and Casadesús Martí**

**Abstract** This chapter analyses the motivations and key factors reported by organizations after successfully implementing a service management system according to the ISO 20000 standard. ISO 20000 (approved in 2005 and revised in 2011) defines the requirements for implementing a standardized service management system, and has a form similar to the pre-existing general quality management standards while adapting its contents to standardize Information Technologies Service Management practices, although there is nothing to prevent it being used in other service management fields. The research is based on a survey that was answered by 105 ISO 20000 certified, Spanish organizations. It enables the profile of these organizations, as well as their main reasons for, and key factors when, implementing the standard, to be characterized. Motives are divided into external and internal factors and, as such, reveal the predominance of external reasons when deciding to implement certification. Furthermore, some key factors in successfully implementing the standard are highlighted, especially in terms of management and staff involvement. Other specific factors have also been analysed: such as the norms with which it is integrated, and the reasons why some organizations abandon the certificate. This research is the first reported in the literature to deal with two fundamental issues related to the implementation of the ISO 20000 standard: the reason behind implementing the standard and what the key factors in its adoption are.

**Keywords** ISO 20000 · ITIL · ITSM · Management standards · Service quality · Service management

C. Santi (✉)
Servei Informàtic, Universitat de Girona, Girona, Spain
e-mail: santi.cots@udg.edu

C. Martí
Departament d'Organització, Gestió Empresarial i, Desenvolupament de Producte, Universitat de Girona, Girona, Spain
e-mail: marti.casadesus@udg.edu

 
I. Heras-Saizarbitoria (ed.), *ISO 9001, ISO 14001, and New Management Standards*, Measuring Operations Performance, https://doi.org/10.1007/978-3-319-65675-5_5

# 1 Introduction

Most of the value delivered by Information Technology (IT) to users takes the form of services (Bitner et al. 2000), whatever definition of value one decides to adopt (Davis et al. 2011). When IT services are produced by organizations, the management of those services becomes a necessity (Radovanovic and Sarac 2011), at least for those who want to deliver them with sustainable quality (Gupta et al. 2005; Stamenkov and Dika 2015).

Some organizations choose a formal and documented approach to their management activity through implementing a formal management system. If the management system follows the requirements of a standard, it could be said that the organization has or owns a standardized management system.

The benefits of standardized management systems have been broadly studied for the most widely-used standards; specifically ISO 9001 and ISO 14001 (Buttle 1997; Casadesús et al. 2001; Gotzamani and Tsiotras 2002; Casadesús and Karapetrovic 2005; Psomas et al. 2011). The service management standard ISO 20000 adopts a similar form to that of pre-existing general quality management standards, but adapts its contents to standardize IT Service Management (ITSM) practices, by defining the requirements to implement such standardized service management systems.

As the standard has rapidly reached a mature level of diffusion (Cots and Casadesús 2015), and in order to gain a deeper understanding of its impact, it would seem relevant to analyse the motivations for, and the key factors in, deploying IT service management systems by organizations that have already obtained ISO 20000 certificates. To this end these aspects are analysed in this chapter.

# 2 Standardisation and ISO 20000

The phenomenon of 'Management Standards' has been widely diffused as exemplified by the widespread use of ISO 9001, aimed at defining the characteristics of a standardized quality management system, and of ISO 14001 which focuses on environmental management systems (Marimon et al. 2006; Psomas 2013; Castka and Corbett 2015). Beyond general purpose management standards represented by, but not limited to, the standards mentioned above, a growing collection of sectorial or specific standards aimed at guiding more specific aspects of management, standards such as ISO 26001 (Corporate Responsibility Management System), ISO 50001 (Energy Management System), etc., have been developed.

Additionally, a common threat of most management standards is that they are auditable. This auditability characteristic means that anyone qualified can check an actual management system against the standard's requirements by using a procedure called 'audit', and, in need be, declare compliance with the standard. Thus, independent organizations specialized in carrying out these audits, ruling on compliance and issuing the so-called "certificates" have emerged. It is important to stress how a certificate relies on the prestige of the issuer and that it is only issued when compliance with the standard is complete i.e. certifying "partial" compliance is not possible.

Within this framework, ISO 20000 is the only international standard that specifically defines a service management system. The standard originated in the desire to establish a management model for Information technology (IT) services, meaning it could be referred to, at least in its beginnings, as a sectorial standard. However, there is widespread opinion that ISO 20000 also seeks to define a model that is applicable to many other sectors, to transform itself from a sectorial standard to a specific but multi-sectorial one for management services beyond IT (Cots and Casadesús 2013).

ISO 20000, which was based on the pre-existing British norm BS 15000, was officially approved by ISO in 2005. The latter was adapted to become an international norm using the procedure known as 'Fast-track', which allows the test period to be shortened under certain conditions. Since the first document, ISO 20000 has been progressively extended with the editing of different documents or parts. Of these, only the first establishes the requirements needed for a management system and so is the only one that serves as a basis for a certification audit. Table 1 shows the current set of documents that conforms the whole ISO 20000 standard.

## 3 ISO 20000 in Spain and Research Questions

Historically, Spain has been a leader in implementing management standards, such as ISO 9001 and ISO 14001 (Casadesús et al. 2001, 2008) and, in a similar way, ISO 20000 (Cots and Casadesús 2015). Given that this chapter is based on

**Table 1** Current documents of ISO/IEC 20000 series

| Document | Name |
|---|---|
| ISO/IEC 20000-1:2011 | Part 1: Service management system requirements |
| ISO/IEC 20000-2:2012 | Part 2: Guidance on the application of service management systems |
| ISO/IEC 20000-3:2012 | Part 3: Guidance on scope definition and applicability of ISO/IEC 20000-1 |
| ISO/IEC TR 20000-4:2010 | Part 4: Process reference model |
| ISO/IEC TR 20000-5:2013 | Part 5: Exemplar implementation plan for ISO/IEC 20000-1 |
| ISO/IEC TR 20000-9:2015 | Part 9: Guidance on the application of ISO/IEC 20000-1 to cloud services |
| ISO/IEC TR 20000-10:2015 | Part 10: Concepts and terminology |
| ISO/IEC TR 20000-11:2015 | Part 11: Guidance on the relationship between ISO/IEC 20000-1:2011 and service management frameworks: ITIL® |

the analysis of certified companies in Spain, it is important to briefly set in context the specific characteristics of this country in relation to the standard under study.

First, the fact that in Spain, like in other countries, there is a company whose various roles include carrying out functions of normalisation (the creation of standards) must be taken into account. This company represents the Spanish position and proposals in relation to ISO, and at the same time acts as a certifying entity in the market. In the case of ISO 20000, it can be easily estimated that the quota of certifications issued by the aforementioned company is more than 50% of the total issued in Spain. Another important factor, which could differentiate Spain from other countries, is that for some years there were official programmes that facilitated obtaining certification, which could have encouraged companies who would otherwise not have opted for achieving certification to do so. In all, it seems that a thorough study of the motivations and key factors in implementing ISO 20000 in Spain is sufficiently representative and can be extrapolated to other markets once the differential characteristics have been considered.

It would seem even more salient, then, to pinpoint and highlight the fact that at the time this study was undertaken, the standard had been in force for just eight years, during which time its diffusion had been rapid but limited (Cots and Casadesús 2015).

Regarding the motivations and key factors in implementing a specific standard like the ISO 20000, even though they are the same of mature global standards, the degree of influence of each of them may be different for that of this young and specific standard must be considered. Furthermore, they may even have varied according to the evolution of society itself or to the phenomenon of standardisation. Thus, studying the motivations and key factors in implementing a standard like ISO 20000 could provide information not only on the standard itself, but also on the standardisation of management in general and its evolution.

Finally, one valuable characteristic of some management standards is the possibility of integration or integrated installing (Karapetrovic et al. 2006; Simon et al. 2012) in which a single and holistic management system conforms to several standards by fulfilling all their requirements. As integration has a lot of advantages, ISO is currently in the process of reviewing all its management standards so that they will share a common form, which will make them much easier to integrate and assist integrated audits. While most of the previously cited standards have undergone this revision process and have already taken on the new structure, ISO 20000 will have to wait until its next revision, scheduled for about 2018, to do so. It is within this framework that questions about how this standard is currently integrated, or not, with others that are implemented within the organisation must be asked.

## 4 Empirical Study: Methodology

The main contribution of this chapter is based on field work that aimed to discover the motivations and key factors in implementing ISO 20000 through the experience of certified companies. To do so, the methodology previously tested by Casadesús et al. (2008) is used.

To investigate the perceptions of the organisations that have obtained a certificate based on ISO 20000, it was decided that those responsible for it within the companies should be surveyed. To this end, after having revised the existing literature on ISO 20000, a research framework and specific questionnaire were designed. Only one previous study with a comparable objective and methodology, although rather more limited in scope, was found (Disterer 2012). Existing studies on ISO 9000 and ISO 14000 were also considered to be basic references (Buttle 1997; Corbett et al. 2003; Karapetrovic et al. 2006, 2010).

In order to decide which concepts would be analysed and which questions would be formulated, all of the questions posed in the research of Buttle (1997), Corbett et al. (2003) and Disterer (2012) were systematically gathered. In this way, a relationship between the questions and categories to be analysed regarding motivations, implementing the standard and obtaining a certificate were established. Those that concerned similar, identical or globalised concepts were grouped together to create a new, unified list of questions which, in one way or another, included all of the concepts from the references. A Likert-type 1–5 unipolar scale of categories (Cañadas and Sánchez 1998) was used to collect the answers.

In order to study its integration with other standards, those that were considered to be relevant to ISO 20000 (ISO 9001, ISO 14001, ISO/IEC 27001, ISO 22301, ISO 31000, ISO 38500, COBIT, ITIL) were selected, including the most widely-diffused standards worldwide and also the main ones from the IT sector or this area (security, continuity).

It was decided that the questionnaire would be conducted through the individualised web formula where a personalised link is sent to each participant using a web platform. That way, the status of each answer could be monitored while ensuring that nobody outside the selected population could input questions into the system.

Once an initial questionnaire was designed, a panel of 8 reputable experts was selected to validate the content of both the questionnaire and the platform. Three of these were academics who had carried out similar studies on other standards at some time previous, and the other 5 were experts from itSMF Spain and distinguished members of the sector with excellent knowledge of the norm in question. For the study, the aim was to send the questionnaire to a discerning individual in a position of responsibility (CEOs, CIOs, quality assurance managers, or similar) in each of the companies that had obtained an ISO 20000 certificate in Spain at some point.

Table 2 outlines the data of this study, highlighting the answer ratio of 70% that represents the 105 questionnaires answered out of the 149 sent, which is probably a

**Table 2** Characteristics of the study

| | |
|---|---|
| Date | May–July 2013 (3 months) |
| Population (estimated) | 186 companies |
| Study sample | 149 questionnaires sent |
| Valid answers received | 105 answers received |
| Answer ratio | 70.46% |
| Maximum error (p = q = 0, 5) | ±6.32% |

reflection of the participants' interest in the matter and of our insistence sending reminders by email and making personal phone calls. The maximum error of ±6.32% assuming normality, based on the approximated population and the number of answers with 95% reliability received, confirms the representability of the results.

Finally, note that during the data treatment process (September–October 2013), several of the participants were contacted via email to verify data that, a priori, seemed to be incoherent (such as dates that did not tally). Where participants' answers were able to clarify one of the aforementioned inconsistencies, the register was corrected accordingly. A deeper analysis of the content of the whole research is available in Cots (2014).

## 5 Motivations for Implementing ISO 20000

To find out the level of agreement regarding the possible motives that lead the companies surveyed to seek and obtain ISO 20000 certification, a list of motivations grouped into related concepts and based on the literature was proposed to them.

Figure 1 is a summary of these motivations ordered according to the average of the answers received for each, on a Likert 1–5 scale. The average value for each of the answers gives us a good idea of what the predominant motivation for most organisations were. Observe how only audits have a below average agreement value (low to medium).

Figure 2 is a graphic overview that gives us a deeper understanding of how the answers are distributed. For example, it shows that although user satisfaction and competitive advantage have valuation averages that are virtually the same, in the first there is a higher consensus as nobody expressed no or very low approval, whereas there were some opinions of this kind when participants were asked about competitive advantage as a motivation.

A usual classification of motivations, and one that was used previously by Buttle (1997) to study the motivations for implementing ISO 9000, suggests grouping them into internal and external motivations. Internal motivations are those that are directed towards making development and improved organisation possible, whereas external motivations refer to promotion and marketing, pressure from clients or other entities or increases in market shares, to give some examples. Although, of

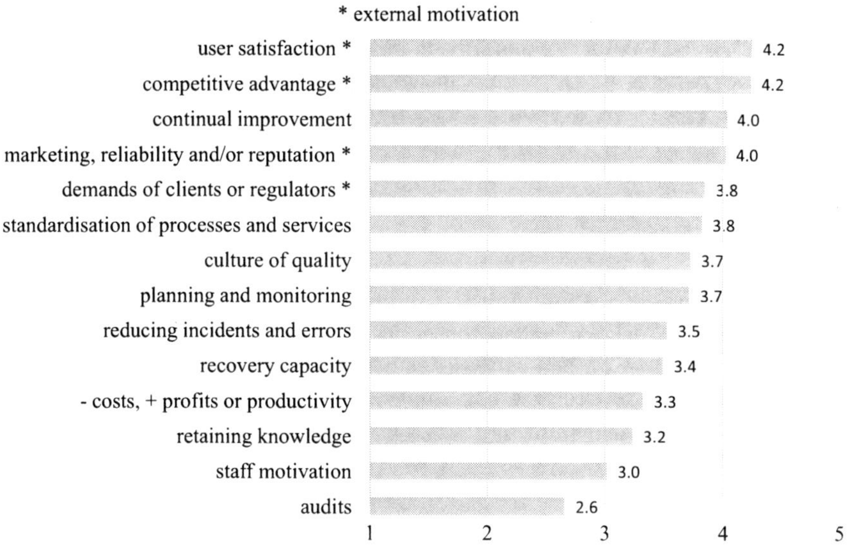


**Fig. 1** Motivations to implement ISO 20000

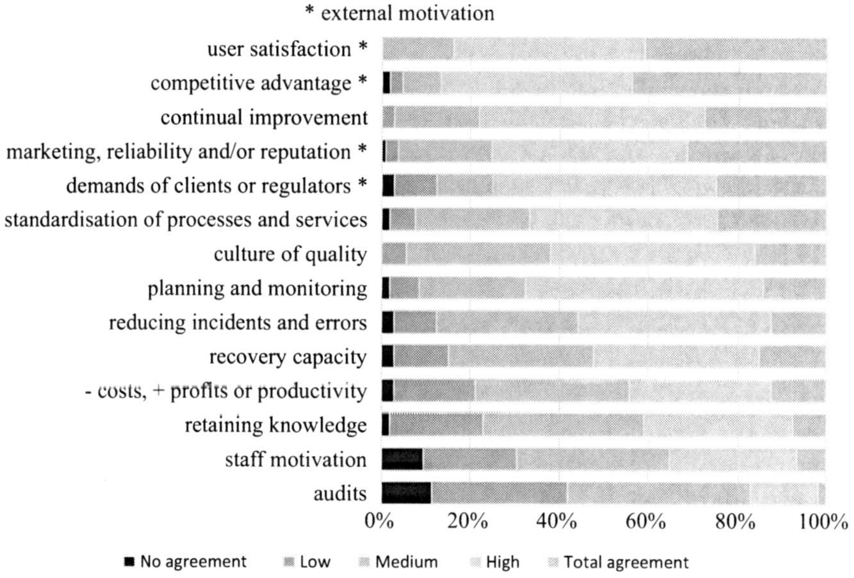


**Fig. 2** Distribution of motivations to implement ISO 20000

course, participants were not informed about this classification, and neither was a specific order followed or grouping made in the questionnaire, the external motivations were marked with a "*" in both figures.

While one type of motivation is as legitimate and respectable as another, it is usual for companies to have their own mix of the two types of motivations. It is also true that the list of internal motivations suggested is longer than the list of external motivations, as it was in the reference studies used and that some studies suggest that internal motivated organizations tend to internalize more the quality standards (Tarí et al. 2013).

Starting the analysis of motivations with the external ones, it can be seen how user satisfaction just has the highest consensus. The very nature of the norm and the services is sufficient to explain why user satisfaction is a key motivation, if not the main one, in taking a decision like implementing a standardised management system for managing the company's services. There is almost the same degree of consensus for competitive advantage. It seems reasonable to assume that the relative youth of the norm at the time of the study fostered a much more accentuated perception of competitive advantage among the first companies to adopt it than there would have been in a more mature market where most of the sector had already implemented it. Thus, if the implementation of the norm becomes generalised, as have enormously successful standards such as ISO 9000, the motivation of competitive advantage would be expected to be progressively substituted by the demands of clients or regulators, which currently features as the last of the external motivations (while still way ahead of most of the internal ones).

In any case, the four affirmations that encompass external motivations feature among those with the highest agreement, allowing us to confirm that, in general, the initial motivations for obtaining ISO 20000 certification are external. At the opposite extreme, the least valued motivations were found to be those linked to staff motivation and especially audits.

## 6 Implementation of ISO 20000: Key Factors

It is pertinent to analyse the key factors involved in successfully implementing the management system and its later external certification. Tacitly identifying obtaining certification with success, as all of those surveyed represent certified companies, we can assume that they have been successful in this area. Thus, their opinion, based on experience, must represent the factors that are key to success.

Figure 3 shows the average opinions of participants with respect to seven success factors. The degree of concordance for all of them is quite high, such that they can all be confirmed to be success factors. Most of the factors have a degree of agreement within the range of 3.5–4, on a scale of 1–5.

Staff, and especially management, involvement stand out with a very high degree of concordance. In fact, it can be seen in the distribution shown in Fig. 4

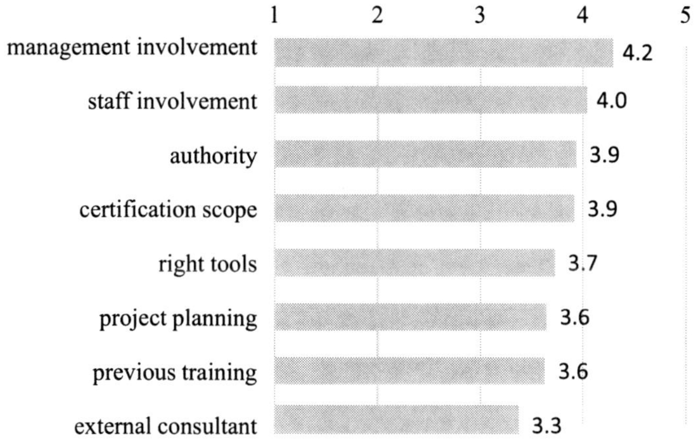


**Fig. 3** Key factors for successful implementation

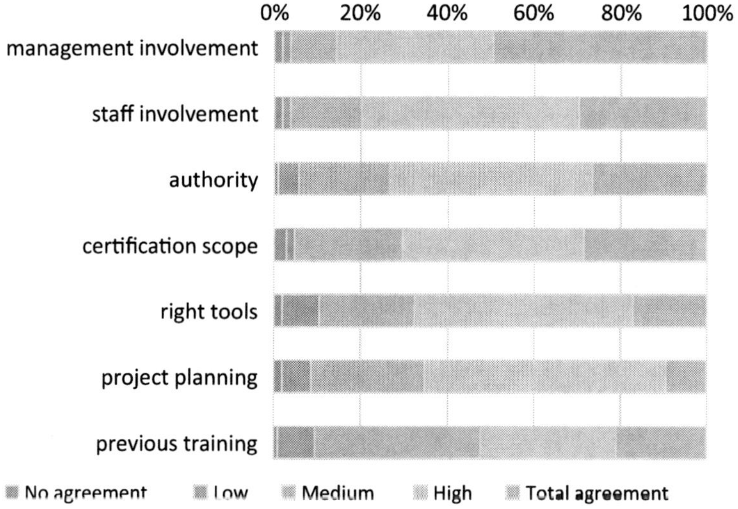


**Fig. 4** Distribution of key factors for successful implementation

how approximately 50% of participants totally agree that management involvement is a key factor.

Regarding the need for a good consultant to undertake the project, even though on average it appears to be a less determinate factor than the rest, it is important to note the dispersion of the answers here. The fact that there are certain companies that successfully achieve certification without turning to external consultants for assistance makes the average for this success factor quite low. On the other hand, many participants either totally agreed or highly agreed with this factor, more so in fact than for previous training. It must be pointed out that in the open field of the questionnaire a couple of participants highlighted the role of consultants or auditors as key success factors, thus recognising the importance and value of these figures in their specific cases.

It can also be observed that staff involvement is considered to be key to success in this type of project, despite staff motivation not being one of the objectives that stand out for obtaining certification.

## 7 Integrating ISO 20000 with Other Standardised Management Systems (SMS)

The joint and simultaneous use of different systems in a single management system is generically known as integration or integrated implementation. Additional benefits are normally obtained from this type of integrated system thanks to the synergies and optimisation of the management itself (Karapetrovic et al. 2006; Bernardo 2012). The use of integrated management systems is a form of efficient organisation that is highly beneficial to those who use them (Casadesús et al. 2011).

The capacity to integrate with certain ease is a characteristic required by all standards. It adds value to some standardised management systems and is actively pursued by organisations like ISO who facilitate integration with each successive version of the norms. Consequently, many standards increasingly tend to use more common and/or compatible structures as they are revised and/or reedited.

At the same time, integration itself, or the knowledge and eventual use of other standards, can indirectly serve to characterise different organisations. There are organisations that are more clearly inclined towards standardised management, that feel comfortable following this type of norm and consider them to be part of their arsenal, while other companies do not display this tendency or have only recently started up and are in the process of maturation. Evidently, not all of the standards offer the same value, nor are they applicable to all organisations. Each potential user must be aware of what standards are available and decide which of them will provide solutions that meet their individual needs.

Thus, in a scenario where different standards, reference frameworks, etc., could be used, organisations can take different stances. For the purposes of this chapter, it was decided that these would be graded from one end of the scale whereby the organisation ignores the existence of standards (either because they do not know about them or as a conscious decision) or simply considers that their area of application has no place for them, to the other end of the scale whereby the organisation decides to make fully integrated use of them and obtain certification in the same standardised management system, as has been done for ISO 20000.

Somewhere on the scale a position can be found where the norm is used as a reference (that is, it is taken into consideration in some way). Another position can be found where the norm is used formally but without certification, another where a standardised management system has been implemented and separately certified, another where the norm is formally integrated with the ISO 20000 management system, but no certification is sought for the additional norm, and finally, another where both (or more) are integrated and certified together.

To find out the state of integration, a list of norms and common standards in the sector was proposed: ISO 9001, ISO 14001, ISO/IEC 27001, ISO 22301, ISO 31000, ISO/IEC 38500, COBIT and ITIL. However, not all of the standards proposed in the study are certifiable. Answers for all of the categories of integration and standards were admitted, although some of the possibilities were not formally possible. In fact, the number of answers that affirmed formally unviable degrees of integration for some standards is significant and requires an analysis of the specific causes. Apart from the odd erroneous answer, some participants may not have been completely clear about the concept of integration or the market (consultants) may even have support a specific idea of the non-certifiable standards, which would have contributed to this confusion.

In any case, leaving aside the 'false integrations' it can be affirmed, as illustrated in Fig. 5, that there are a large number of organisations that use ISO 9001, ISO 27001 and, to a lesser degree, ISO 14001 in an integrated way with ISO 20000, or who maintain separate certificates. The use of ITIL is also very widespread, as was expected given its popularity in the sector and its close relationship with ISO 20000.

Finally, in an attempt to find out which norms are used in the sector and, at the same time, to better identify companies with a normal experience, participants were asked to identify which other standards they use. These are shown in Table 3.

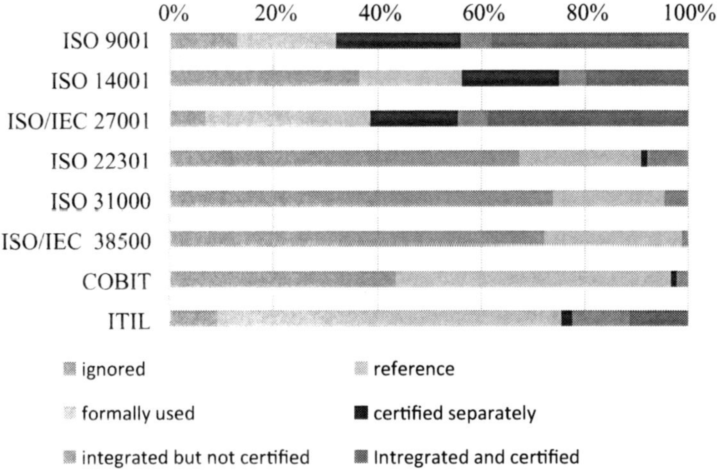


**Fig. 5** Integrations with ISO 20000

**Table 3** Other standards used

| Standard or norm | Answers |
|---|---|
| ISO 15504 | 7 |
| CMMI | 7 |
| UNE 166002 | 3 |
| EFQM | 2 |
| UNE-EN ISO 14006 | 1 |
| PECAL 2110 y 2210 | 1 |
| OHSAS 18001 | 1 |
| PMBOK | 1 |
| MOF | 1 |

As can be seen, the standards most often mentioned correspond to the field of software development, while among the special cases some very specific standards or norms can be found, such as military regulations.

## 8 Conclusions

First, it must be stated that a broad vision of the motivations and key factors in implementing ISO 20000 has been offered in this study, by means of comparing it with other standards.

A large number of answers were received, and, because of the population that was sampled for the survey, the data are highly reliable. The first conclusion concerns the motivations for implementing ISO 20000. It can be concluded that external motivations are key in the decision to initiate the process of implementing a standardised service management system. In other words, motivations related to service users, image and competition are primary.

Regarding the implementation process through which those who manage to achieve certification pass, the importance of the human factor, in the form of management, and especially staff, involvement must be mentioned. The strong relationship between ISO 20000 and ITIL also seems to be foremost for those that implement the standard, although this could be seen as a weakness when ISO 20000 is implemented in other sectors beyond IT.

This relationship with other standards becomes particularly pertinent when we see that more than half of the companies with ISO 20000 certification also have ISO 9001 and/or ISO 27001 certification, mostly in an integrated management system. 35% have ISO 14001 certification, half of them integrated. Clearly, this high percentage of companies with other certifications allows us to define two types of organisations according to their experience or relationship with other standardised management systems. Without doubt, this study is a reflection of a certain moment in time, of a norm that is still very young. Consequently, the findings reported here can serve as a reference for future developments in this field.

In any case, the last conclusion is, in fact, that this is an open area of research and knowledge and it is hoped that this small contribution, along with others, will serve as a springboard for future advances that will allow us to continue to widen our knowledge of the field of service management and standardisation.

## References


Bernardo M (2012) Quality management and beyond: the current situation and future perpectives. Documenta Universitaria, Girona

Bitner MJ, Brown SW, Meuter ML (2000) Technology infusion in service encounters. J Acad Mark Sci 28(1):138–149

Buttle F (1997) ISO 9000: marketing motivations and benefits. Int J Qual Reliab Manage 14 (9):936–947

Cañadas I, Sánchez A (1998) Categorías de respuesta en escalas tipo Likert. Psicothema 10:623–631

Casadesús M, Karapetrovic S (2005) The erosion of ISO 9000 benefits: a temporal study. Int J Qual Reliab Manage 22(2):120–136

Casadesús M, Giménez G, Heras I (2001) Benefits of ISO 9000 implementation in Spanish industry. Eur Bus Rev 13(6):327–336

Casadesús M, Heras I, Karapetrovic S (2008) The 9000 with the 9000: an analysis of the impact of the ISO 9000 standard in Catalonia, CIDEM, Barcelona

Casadesús M, Karapetrovic S, Heras I (2011) Synergies in standardized management systems: some empirical evidence. TQM J 23(1):73–86

Castka P, Corbett CJ (2015) Management systems standards: diffusion, impact and governance of ISO 9000, ISO 14000, and other management standards. Found Trends Technol Inf Oper Manage 7(3–4):161–379

Corbett CJ, Luca AM, Pan J-N (2003) Global perspectives on global standards. In: ISO management systems. No. January–February, pp 31–40

Cots S (2014) Impacto de ISO 20000. Un estudio empírico, Documenta Universitaria, Girona

Cots S, Casadesús M (2013) Implementing ISO 20000: proposals from learned lessons. TMQ Tech Methodol Qual 4:12–31

Cots S, Casadesús M (2015) Exploring the service management standard ISO 20000. Total Qual Manage Bus Excell 26(5–6):515–533

Davis MM, Spohrer JC, Maglio PP (2011) How technology is changing the design and delivery of services. Oper Manage Res 4(1–2):1–5

Disterer G (2012) Why firms seek ISO 20000 certification—a study of ISO 20000 adoption. In: ECIS 2012 proceedings, Barcelona, Spain, Paper 31

Gotzamani KD, Tsiotras GD (2002) The true motives behind ISO 9000 certification: their effect on the overall certification benefits and long term contribution towards TQM. Int J Qual Reliab Manage 19(2):151–169

Gupta A, McDaniel JC, Herath SK (2005) Quality management in service firms: sustaining structures of total quality service. Manage Serv Qual 15(4):389–402

Karapetrovic S, Casadesús M, Heras I (2006) Dynamics and integration of standarizaed systems. Documenta Universitaria, Girona

Karapetrovic S, Casadesús M, Heras I (2010) What happened to the ISO 9000 lustre? An eight-year study. Total Qual Manage Bus Excell 21(3):245–267

Marimon F, Casadesús M, Heras I (2006) ISO 9000 and ISO 14000 standards: an international diffusion model. Int J Oper Prod Manage 26(2):141–165

Psomas EL (2013) The impact of ISO 9001 effectiveness on the performance of service companies. Manage Serv Qual 23(2):149–164

Psomas EL, Fotopoulos CV, Kafetzopoulos DP (2011) Motives, difficulties and benefits in implementing the ISO 14001 Environmental Management System. Manage Environ Qual Int J 22(4):502–521

Radovanovic D, Sarac M (2011) Necessity of IT service management and IT governance. In: MIPRO, 2011 proceedings of the 34th international convention, Opativa, pp 1430–1433

Simon A, Karapetrovic S, Casadesús M (2012) Difficulties and benefits of integrated management systems. Ind Manage Data Syst 112(5):828–846

Stamenkov G, Dika Z (2015) A sustainable e-service quality model. J Serv Theory Pract 25 (4):414–442

Tarí JJ, Heras I, Pereira J (2013) Internalization of quality management in service organizations. Manage Serv Qual 23(6):1

# The Internalization of a Sectorial Standard for Quality Management: A Qualitative Analysis in Tourism

**Juan-José Tarí, José F. Molina-Azorín, Jorge Pereira-Moliner and María D. López-Gamero**

**Abstract** The aim of this work is to analyze the motives for adopting a quality standard and the internalization process in tourism organizations with a sectorial quality certificate using a qualitative study. The results show, first, that the motives for adopting quality standards can be internal and external (for example, quality culture, management commitment, employee management, and the pressure from customers and society). Second, the following enablers are essential for the internalization process: management commitment, initial and ongoing training for employees, application of documentation to daily routines and follow-up.

**Keywords** Internalization · Quality certificate · Tourism industry · Qualitative analysis

## 1 Introduction

The adoption of quality standards has been widely examined in the operational field, but few works have focused on the internalization process of these standards. Most works that have examined quality standards have measured their adoption using a binary variable (Sharma 2005; Bayo-Moriones et al. 2011), that is, they have considered whether the organization had a quality certificate, thus assuming a homogeneous implementation of quality. A few studies have gone further and have analyzed heterogeneous adoption (internalization) (Naveh and Marcus 2005; Boiral

J-.J. Tarí (✉) · J.F. Molina-Azorín · J. Pereira-Moliner · M.D. López-Gamero
Department of Business Management, University of Alicante, Alicante, Spain
e-mail: jj.tari@ua.es

J.F. Molina-Azorín
e-mail: jf.molina@ua.es

J. Pereira-Moliner
e-mail: jorge.pereira@ua.es

M.D. López-Gamero
e-mail: md.lopez@ua.es



I. Heras-Saizarbitoria (ed.), *ISO 9001, ISO 14001, and New Management Standards*, Measuring Operations Performance, https://doi.org/10.1007/978-3-319-65675-5_6

2011; Ataseven et al. 2014). In other words, they have considered that organizations may implement quality standards to a greater or lesser extent.

There are few studies analyzing the process of heterogeneous implementation of quality standards, and qualitative studies are in a minority (Boiral 2003, 2011; Heras-Saizarbitoria 2011). In addition, most studies of the internalization of quality standards have examined manufacturing companies. More qualitative studies are needed in other sectors to produce a more detailed analysis of how heterogeneous adoption takes place (internalization) and why. The last question addresses the motives for certification and the process of adopting the quality standard.

The aim of this work is to analyze the motives for adopting a quality standard and the internalization process in the case of tourism organizations with a sectorial quality certificate using a qualitative analysis. This sectorial quality certificate has been created based on the ISO 9001 standard as a new management standard for tourism organizations. This sectorial quality standard is similar to ISO 9001 because it includes the ISO 9001 requirements, but there are some differences: (a) the ISO 9001 standard does not set service criteria and the sectorial standard for tourism includes all the service quality specifications within the standard itself, and (b) the ISO 9001 certificate can be applied to any industry or organization, while the sectorial standard for tourism applies only to the tourism industry in Spain.

This sectorial standard for quality management was promoted in Spain in the 1990s. More specifically, a number of policies were defined by the government in order to improve the quality of the tourism industry. On this basis, the Spanish Institute for Tourism Quality (ICTE) was created, together with the standards for tourist quality. In this way, in addition to the ISO 9001 standard, which is the most widely known quality standard, there is another quality standard in Spain which applies only to the tourism industry, which also allows the firm to obtain a quality certificate: the Q certificate.

## 2 Literature Review

Quality standards can be a commercial tool and/or a way to adopt an internal management system (Boiral and Roy 2007). Institutional theory and resource-based view can explain these different points of view. Institutional factors (DiMaggio and Powell 1983) can influence the adoption of certain management practices (Meyer and Rowan 1977), such as quality standards (Nair and Prajogo 2009; Heras-Saizarbitoria and Boiral 2013). These factors can lead to a superficial adoption of the quality standard (Martínez-Costa et al. 2008; Nair and Prajogo 2009) to satisfy external pressures (e.g. society, governments and/or customers) to have the quality certificate (Singels et al. 2001; Rubio-Andrada et al. 2011), to imitate those companies implementing quality systems (Nair and Prajogo 2009), and/or to improve corporate image in the market (Jones et al. 1997; Prajogo 2011).

The resource-based view suggests that organizations focusing on internal motives are more motivated to adopt a quality system for continuous improvement.

If organizations win a quality certificate for internal reasons, they may create capabilities which are harder for their competitors to imitate (Martínez-Costa et al. 2008; Prajogo 2011) that can lead to a higher level of quality standard implementation.

Accordingly, internal and/or external motives for certification influence the way a quality standard is implemented (Boiral and Roy 2007; Nair and Prajogo 2009). Amongst them, internal motives are more important than external ones (Prajogo 2011) and, according to some authors, external reasons have no impact upon internalization. Alongside these reasons for certification (see Table 1 for the most common reasons), other motives suggested by some studies on internalization are the following:

- Quality culture (Briscoe et al. 2005). Those organizations showing a greater interest towards having a quality culture will encounter fewer problems in adopting quality standards.
- Leadership. Leadership can cause the organization to develop more customer-focused practices, to provide more training to employees, and to improve relationships with suppliers (Singh 2008).
- Training. Trained staff finds it easier to integrate the quality standard requirements in their work routines (Naveh and Marcus 2005).
- Frequent audits by customers. These audits may lead the firm to implement the quality standard to a greater extent (Christmann and Taylor 2006), for instance, as a way to ensure improved product and/or service quality.
- Innovation environment. Innovation helps organizations to introduce improvements and, therefore, to adopt quality-related practices (Briscoe et al. 2005).
- System coordination with suppliers and customers. An improvement of these relationships facilitates the level of implementation of the requirements of the standard (Briscoe et al. 2005; Naveh and Marcus 2005).

These ideas indicate that other motives, different from those most commonly studied previously (see Table 1), such as quality culture, leadership, training, innovation and customer and supplier relationships, may influence also quality standard adoption.

**Table 1** Most common reasons for certification according to the literature

| Motives | Studies |
|---|---|
| *Internal motives*<br>Service/product quality, improved systematization, errors and cost reduction, improved efficiency | Brown et al. (1998), Huarng et al. (1999), Arauz and Suzuki (2004), Allur et al. (2014) and Ataseven et al. (2014) |
| *External motives*<br>Improved image, competitive advantage, customer requirements, selling in other markets | |

In relation to the implementation process, on the basis of the qualitative studies (Boiral 2003, 2011; Heras-Saizarbitoria 2011) and the quantitative studies (Arauz and Suzuki 2004; Briscoe et al. 2005; Naveh and Marcus 2005; Jang and Lin 2008; Singh 2008; Nair and Prajogo 2009; Prajogo 2011; Prajogo et al. 2012; Psomas et al. 2013; Tarí et al. 2013; Allur et al. 2014; Ataseven et al. 2014) about internalization, some ideas may be put forward about the aspects that organizations should consider important for internalization:

- Integration of the quality standard requirements in the organization's day-to-day practices, for example, when the documents are used in day-to-day practices and employees have the documents available in the workplace.
- The documents are updated regularly and employees participate in the creation/modification of documents.
- Employees are trained in quality and ISO 9000 standards.
- Managers value and are committed to internal audits and the participation of employees in internal audits, and external audits are not prepared at the last moment.
- Managers use general data to find solutions and improvements, that is, the system makes it possible to identify improvement opportunities in order to innovate and introduce new practices.

On this basis, the following research questions are formulated:

RQ 1: Which are the internal and external motives influencing internalization in tourism organizations?
RQ 2: How is internalization developed in tourism organizations?

# 3 Methods

An empirical exploratory study is used to respond to the two research questions in order to contextualize in the tourism industry the reasons for the internalization of the quality standard requirements and their internalization level.

## *3.1 Data Collection*

Data collection was performed through detailed semi-structured interviews. A total of nine interviews were held, two with persons responsible for tourism associations (A1 and A2), one with a quality consultant (QC), three with three hotel managers (H1, H2 and H3), one with a tourism information office manager (TI), one with a beach manager (BE) and one with a restaurant manager (RE). Data confidentiality and respondent anonymity were guaranteed in order to increase confidence and reduce the possibility of respondents attempting to take on the role of "good subjects', showing the interviewers that they thought what the interviewers wanted to hear (Celsi et al. 1993). McCracken (1988) indicates that such a small sample may be suitable for qualitative studies.

The respondents were located by telephone and the interviews were held at the respondent's workplace, with an average duration of one and a half hours each. The interviews were recorded with the consent of the respondent and transcribed. The information was completed with data from websites and direct observation. The interview transcripts were checked by the respondents in order to ensure their accuracy. The quotations that appear below are from the transcripts. Respondents were asked about the reasons influencing the internalization of the quality standard requirements, and about the internalization process of such requirements.

### 3.2 Data Analysis

A content analysis was performed based on the transcripts from the interviews. The contents of each interview were analyzed immediately after the interview and were used as the basis in order to explore the resulting ideas on subsequent interviews on the variables in the study.

In order to reduce potential bias, multiple methods and data sources were used. The data from the interview were triangulated through the analysis of the qualitative contents of public corporate documents, such as annual reports, quality documents from the organizations, website information, press publication and direct observation.

The interview transcripts were examined through the analysis of the ideas expressed by the respondents. After each interview, a summary was prepared (Miles and Huberman 1984) in order to emphasize new ideas, variables and other relevant issues which were to be taken into account in subsequent interviews. Data analysis generated a list of reasons and a list of enablers for internalization.

## 4 Results

### 4.1 Evaluation of RQ 1: Which are the Internal and External Motives Influencing Internalization?

Initially, the following question was asked: *In your opinion, which factors would contribute to the quality standard requirements being internalized by all employees?* The respondents stated that the reasons for success in the implementation, maintenance and adoption of the quality standard were the quality culture existing in the firm, management and employee commitment, continuous training and communication and internal follow-up:

A1: "The first and foremost factor is having the involvement and the support of top management".
A2: "Training, management knowledge, indicators, team involvement and complicity (coordination among departments), team work".
OT: "Firstly, I'd mention the management's involvement by providing those requirements included in the standard, and another very important point is the employees being committed to quality. If the management is not committed, employees eventually lose motivation and it is very difficult to reach the objectives".
H3: "Follow-up is a must for improvement. We hold two monthly meetings, and then day-to-day contact and control by the person responsible for day-to-day records".
RE: "We have obtained greater control in issues related to hygiene, cooking and temperature. We have also improved in correct goods storage (from the oldest to the most recent goods)".

Regarding management commitment (*Which aspect from your corporate culture have helped to internalize the standard?*), the respondents answered that leadership is a basic element in order to create such culture. The respondents stated the following:

H1: "Belonging to an international firm has allowed us to have a quality-supportive culture".
H2: "The fact that the owner of the firm himself is highly committed and believes it necessary that all the quality procedures are followed. He transmits his views to the whole organization".
H3: "Top-down example (the employees have to see that the standard is applied). If I prepare a checklist in order to detect problems and employees fill it in, but the problems are not solved, then the employee will say: Now, what's going on here…".

The above comments show that management commitment is a basic motive to facilitate internalization and prove the usefulness of the documents that must be filled in. This facilitates commitment with employees.

With regard to training (*Which training has been provided on the standard to managers and employees? ¿Which training (quality) programmes exist for all levels in the firm?*), the respondent firms provide continuous training to their employees:

H2: "Training for managers and heads of department; and talks to employees (in each department). Each employee joining the firm must be familiar with all the procedure handbooks in his department. We also have an annual training scheme".
PL: "They are provided with the quality policy, the procedures, new documents, through information talks, […]. Workers contribute ideas from the work reports

they prepare (we try to solve incidents as soon as possible). There is a "remarks" section where they indicate how the development of an activity may be improved in their department or elsewhere".
RE: "Information talks are provided every three months, together with the internal auditor, where employees may contribute their opinions and ideas".

In order to analyze other motives pointed out in the literature, other questions are asked concerning customers, suppliers and society. With regard to customer perception (*Do you think that customer perception has influenced the implementation and internalization of the requirements of the standard?*), the respondents indicated that customer perception has had an impact on the implementation of the standard requirements. Respondents stated that customers do not actually demand a certificate, but their opinions and their perceptions do matter. This idea, mentioned by managers, is supported by A2:

A2: "I don't think so. Customers may provide hints to improve your services if you want to listen to them, but not as a requirement of the standard. The pressure is rather towards service improvement, but not because the quality standard applies".

As for suppliers' perception (*Do you think that suppliers' perception has had an influence on the implementation and the internalization of the requirements of the standard?*), respondents state that suppliers have had no influence upon certification. According to the respondents, it has not been important for certification (their answer was a clear "no"), although such suppliers do participate in the system, because it must fit the requirements set by the organization.

Concerning societal perception (*Do you think societal perception has led to a greater or lower extent of implementation and internalization of the requirements of the standard?*), in general respondents believe that there are no pressures from society towards the implementation of quality (many of them answer "no"), but there is some concern for the environment.

These answers show how external motives (customer requirements, environmental concern and even societal pressures) may have an impact on the internalization of the Q tourism quality certificate. However, the answers show that internal motives (for instance, culture and management and employee commitment) are more important.

## 4.2 Evaluation of RQ 2: How is Internalization Developed?

The first questions tackling this second research question concerned the involvement of management and employees in the internalization process (*What was the employee participation in the implementation and internalization of the standard? ¿What was the participation of middle management? What was the employee participation in the day-to-day use of the standard?*). Employees and the management participate in compliance with the quality standard. How? Respondents have said that there is employee participation in the implementation of the standard, mainly through participation in training and compliance with the documents written. Respondents stated as follows:

H1: "Employees have received training in new work systems and procedures, as they are the ones who eventually perform the processes".
H2: "Everybody has to do what is contained in the procedures, which in turn, is what is done in the firm".

One of the respondents pointed out that, before certification, they already followed a number of service quality standards because they worked with international customers demanding specific requirements, which led them at the time to document process procedures (for instance, in kitchens, concerning fires, etc.). In this respect, respondents told us that the employees are the ones who daily apply the requirements of the standard:

H3: "The employees apply the documents to day-to-day practice [...] If there is a checklist for floors, restaurant, etc. people fill it in [...] The documents are in the workplace and the employees use them".
PL: "The documents are revised by the management and by administrative staff. These documents are handed into the person in charge. The records are applied by the employees in all the beaches".

In addition to employee commitment, managerial involvement in the implementation and development of the system is also a key factor. All the respondents mentioned the important role of the management, which is proved when the management participates in defining the documents, receives and provides training and communicates the process to the employees. These ideas are clearly supported by the respondents from the two associations who, like other respondents, emphasize the role of management involvement in improvement activities and in internal follow-up activities:

H2: "Once a month, the operation of each department is analyzed with the manager, in order to see the evolution of the indicators (follow-up) and if there is any deviation from what is required, it is corrected".

These opinions indicate that another key element of internalization is internal follow-up. A more advanced internalization implies a continuous follow-up of the system, and therefore, being prepared for audits. In this respect, when they were asked how the audits are prepared (*How are audits prepared? At the last stage? Who takes part?*), respondents answered that they were not prepared at the last moment, as proved by the follow-up carried out and the daily application of and compliance with the documentation.

H2: "If you always work with the quality standard, preparing the audit is very easy, because everything is constantly updated. Preparation is a general review, thanks to the continuous follow-up we do."

This indicates that, if the requirements of the quality standard are internalized, it is easy to move on to the audit. However, if the internalization is not a real one, it may usually happen that they are prepared at the last stage:

A1: "In those cases in which the organization has implemented the system with merely commercial purchases, without having internalized a change in the organization regarding its corporate culture, they find, to their surprise, that they have to work during the last week before the audit in order to scrape through the audit."

This follow-up makes it possible to introduce changes or improvements. When asked which changes or improvements have been implemented as a result of the introduction of the standard (*Which changes and improvements have been implemented as a result of the implementation and internalization of the standard?*), the respondents have indicated that the changes or improvements result from the follow-up and the analysis of the information (for instance, of employees' suggestions or indicators) and from customer feedback.

These ideas show that tourism organizations may internalize the quality requirements, first through management commitment, and then by training employees so that they apply the documents to their daily activities. On this basis, periodic controls may be carried out and the management can perform an internal follow-up, for instance, checking record compliance or analyzing information from customers and employees' suggestions. When this internal follow-up is a continued one, the organization may find it easier to introduce continued improvement or changes.

At this stage, a question arose during the interview regarding the barriers to quality implementation (*Which are the main barriers your organization found during the implementation of the quality standards?*). Some of the drawbacks pointed out by the respondents are the documents which must be filled in, resistance to change, lack of commitment and the investment required.

## 4.3 Summary of Motives and Enablers

Based on these ideas, the internal and external motives that facilitate internalization and the key aspects in the internalization process are those shown in Table 2.

**Table 2** Summary of motives and key issues for internalization

| *Motives* |
|---|
| *Internal*: quality culture, leadership, employee commitment, training and follow-up<br>*External*: customer requirements, environmental concerns to a lesser extent, and societal pressure to some extent |
| *Key aspects (enablers)* |
| Management commitment, training, application of document in daily routines, follow-up (indicators, suggestions by employees, etc.), improvements |

## 5 Conclusions

The results obtained from this qualitative study indicate that certified firms with a sectorial standard for quality management, similar to ISO 9001, seek certification for internal and external motives (RQ 1). The internal motives include quality culture and management commitment (for instance, receiving training, communicating with employees, performing follow-up, etc.) and employee commitment (for instance, through training, filling in the records during day-to-day practice, etc.). Alongside these internal reasons, external motives (such as customer pressure, and society to a very small extent) can also have an influence on internalization. These results support the theory that points out that internal and external reasons have an influence on the degree of ISO 9001 adoption and that internal reasons are more important (Boiral and Roy 2007; Prajogo 2011), and extends these results to the case of a sectorial standard for quality management in tourism.

In order to internalize the requirements of the Q certificate (RQ 2), a fundamental requirement is that the management must participate in the drafting and approval of the documents, of initial and continuous training for employees, and that the management perform a follow-up of compliance with the quality requirements through internal meetings, record verification, indicator verification and employees' ideas. In this way, managerial and employee commitment, communication and training may create a quality culture leading to daily compliance with the requirements. In this process, the employees receive information and training making them familiar with their activities and enabling them to comply with the requirements of the quality standard. On this basis, the management may perform an internal follow-up which may lead to the identification of opportunities for improvement, and therefore, to advance towards a greater level of internalization. When progress is made towards continuous improvement, the level of internalization increases. In this respect, it is very important that quality managers are aware which documents they need and which are not required in order to standardize their work routines and improve efficiency without creating excessive bureaucracy or paperwork, which is one of the barriers pointed out. Through communication, training and follow-up, it is possible to reduce resistance to change, which is also mentioned as an important barrier.

This work contributes by supplementing previous studies on internalization, and pointing out that, in addition to the most commonly studied motives for certification (Arauz and Suzuki 2004; Allur et al. 2014; Ataseven et al. 2014), there are others that also have an impact on internalization. These motives may also be key aspects for internalization, as results have been found suggesting that they can be in turn motives and enablers for internalization. Amongst these enablers, one of them, follow-up, has not been emphasized in previous studies in an explicit way, although implicit mention has been made when discussing the need to introduce improvements.

Finally, the work has limitations. It provides only preliminary results due to the exploratory nature of a qualitative methodology. The results cannot be generalized and can be used to contextualize these issues in tourism organizations. In addition, the organizations belong to a specific geographic area; therefore, future studies could focus on analyzing these issues in other contexts and cultures, and using quantitative studies. Despite these limitations, researchers and managers who are interested in quality standards may increase their understanding on the internalization process of sectorial quality standards supplementing previous knowledge about ISO 9001.

## References


Allur E, Heras-Saizarbitoria I, Casadesús M (2014) Internalization of ISO 9001: a longitudinal survey. Ind Manage Data Syst 114:872–885

Arauz R, Suzuki H (2004) ISO 9000 performance in Japanese industries. Total Qual Manage Bus Excell 15:3–33

Ataseven C, Prajogo DI, Nair A (2014) ISO 9000 internalization and organizational commitment-implications for process improvement and operational performance. IEEE Trans Eng Manage 61:5–17

Bayo-Moriones A, Merino-Díaz-de-Cerio J, Escamilla-de-León SA (2011) The impact of ISO 9000 and EFQM on the use of flexible work practices. Int J Prod Econ 130:33–42

Boiral O (2003) ISO 9000: outside the iron cage. Organ Sci 14:720–737

Boiral O (2011) Managing with ISO systems: lessons from practice. Long Range Plan 14:197–220

Boiral O, Roy MJ (2007) ISO 9000: integration rationales and organizational impacts. Int J Oper Prod Manage 27:226–247

Briscoe JA, Fawcett SE, Todd RH (2005) The implementation and impact of ISO 9000 among small manufacturing enterprises. J Small Bus Manage 43:309–330

Brown A, Van der Wiele T, Loughton K (1998) Smaller enterprises' experiences with ISO 9000. Int J Qual Reliab Manage 15(3):273–285

Celsi RL, Rose RL, Leigh TW (1993) An exploration of high-risk leisure consumption through skydiving. J Consum Res 20(1):1–23

Christmann P, Taylor G (2006) Firm self-regulation through international certificable standards: determinants of symbolic versus substantive implementation. J Int Bus Stud 37:863–878

DiMaggio PJ, Powell W (1983) The iron cage revisited institutional isomorphism and collective rationality in organizational fields. Am Sociol Rev 48:147–160

Heras-Saizarbitoria I (2011) Internalization of ISO 9000: an exploratory study. Ind Manage Data Syst 111:1214–1237

Heras-Saizarbitoria I, Boiral O (2013) ISO 9001 and ISO 14001: towards a research agenda on management system standards. Int J Manage Rev 15:47–65

Huarng F, Horng C, Chen C (1999) A study of ISO 9000 process, motivation and performance. Total Qual Manage 10:1009–1025

Jang W-Y, Lin C-I (2008) An integrated framework for ISO 9000 motivation, depth of ISO implementation and firm performance, the case of Taiwan. J Manuf Technol Manage 19: 194–216

Jones R, Arndt G, Kustin R (1997) ISO 9000 among Australian companies: impact of time and reasons for seeking certification on perceptions of benefits received. Int J Qual Reliab Manage 14:650–660

Martínez-Costa M, Martínez-Lorente A, Choi TY (2008) Simultaneous consideration of TQM and ISO 9000 on performance and motivation: an empirical study of Spanish companies. Int J Prod Econ 113:23–39

McCracken G (1988) The long interview. Sage, Newbury Park

Meyer JW, Rowan B (1977) Institutional organizations: formal structure as myth and ceremony. Am J Sociol 83:340–363

Miles MB, Huberman AM (1984) Qualitative data analysis. Sage, Thousand Oaks

Nair A, Prajogo D (2009) Internalisation of ISO 9000 standards: the antecedent role of functionalist and institutionalist drivers and performance implications. Int J Prod Res 47: 4545–4568

Naveh E, Marcus A (2005) Achieving competitive advantage through implementing a replicable management standard: installing and using ISO 9000. J Oper Manage 24:1–26

Prajogo D, Huo B, Han A (2012) The effects of different aspects of ISO 9000 implementation on key supply chain management practices and operational performance. Supply Chain Manage Int J 17:306–322

Prajogo DI (2011) The roles of firms' motives in affecting the outcomes of ISO 9000 adoption. Int J Oper Prod Manage 31:78–100

Psomas EL, Pantouvakis A, Kafetzopoulos DP (2013) The impact of ISO 9001 effectiveness on the performance of service companies. Manage Serv Qual 23:149–164

Rubio-Andrada L, Alonso-Almeida MM, Rodríguez-Antón JM (2011) Motivations and impacts in the firm and stakeholders of quality certification: evidence from small-and medium-sized service enterprises. Total Qual Manage Bus Excell 22:833–852

Sharma DS (2005) The association between ISO 9000 certification and financial performance. Int J Account 40:151–172

Singels J, Ruël G, van de Water H (2001) ISO 9000 series certification and performance. Int J Qual Reliab Manage 18(1):62–75

Singh PJ (2008) Empirical assessment of ISO 9000 related management practices and performance relationships. Int J Prod Econ 113:40–59

Tarí JJ, Heras-Saizarbitoria I, Pereira-Moliner J (2013) Internalization of quality management in service organizations. Manage Serv Qual 23:456–473

# Drivers, Obstacles and Benefits of the Adoption of SA8000: A Survey in Italian Companies

**Iñaki Heras-Saizarbitoria, Giovanna Talamazzi, Francesco Testa and Olivier Boiral**

**Abstract** Several CSR certifiable meta-standards have been launched in recent decades but one of the most popular and widespread has been the SA8000 standard, as the ISO 26000 launched by ISO is not suitable for certification purposes (Hahn in Bus Strategy Environ 22:442–455, 2013). Indeed, in the scholarly literature, SA8000 has been considered to be one of the best CSR initiatives, as it is intended to institutionalize business ethics through standardization (Gilbert and Rasche in Bus Ethics Q 187–216, 2007). Nevertheless, the adoption of this CSR standard has not been widely studied compared to the cases of ISO 9001, ISO 14001 or EMAS. In order to fill this gap in the literature, the aim of the present chapter is to shed light on the adoption of the SA8000 meta-standards by Italian companies. The empirical study focused on the Italian case because it seems paradoxical; Italy is the country in the world with the highest intensity of SA8000 certification in relative terms, even though this CSR standard was targeted at developing countries with labour intensive industries. The work analyzes the motivations, obstacles and benefits of the adoption of the SA8000 standard using primary data obtained from a survey of 130 Italian organizations. The article contributes to the empirical literature on the

I. Heras-Saizarbitoria (✉) · G. Talamazzi
Department of Management, Faculty of Economy and Business,
University of the Basque Country UPV/EHU, San Sebastian, Spain
e-mail: iheras@ehu.es

G. Talamazzi
e-mail: giovanna.talamazzi@hotmail.com

F. Testa
Istituto Di Management, Sant'Anna School of Pisa, Pisa, Italy
e-mail: francesco.testa@sssup.it

O. Boiral
Département de Management, Faculté Des Sciences de L'administration,
Université Laval, Quebec City, QC, Canada
e-mail: Olivier.Boiral@mng.ulaval.ca

 
I. Heras-Saizarbitoria (ed.), *ISO 9001, ISO 14001, and New Management Standards*,
Measuring Operations Performance, https://doi.org/10.1007/978-3-319-65675-5_7

adoption of CSR standards. Implications for managers, policy makers and other stakeholders are discussed, together with avenues for further research.

**Keywords** CSR standards · SA8000 · Meta-standards · Adoption · Survey · Italy

# 1 Introduction

Management system standards, also called meta-standards, have been adopted by an increasing number of organizations across the world (Heras-Saizarbitoria and Boiral 2013; Boiral et al. 2017a). The development of management standards now encompasses a very wide range of aspects of business activity, such as quality management (e.g. ISO 9001), environmental management (e.g. ISO 14001), the prevention of occupational hazards and the provision of health and safety regulations in the work-place (e.g. OHSAS 18000), and corporate social responsibility (e.g. SA8000). These standards tend to use quite similar methodology with regard to their creation, structure, implementation process and monitoring by a third party, a trend that was established by the successful ISO 9001 and ISO 14001 standards. These meta-standards are voluntary codes, guidelines or processes used by organizations to formalize, systematize, and legitimize a very diverse set of managerial activities or tasks (Boiral and Heras-Saizarbitoria 2015). Most meta-standards do not refer to compliance with an objective or with a particular result. For instance, ISO 14001 does not fix environmental goals or environmental targets to be achieved (e.g. reduction of greenhouse gas emissions, recycling, and energy consumption), as a result of the adoption and possible certification of this standard. Rather, ISO 14001 defines the procedural requirements concerning the type of policy, plans, organizational practices and control mechanism to be adopted by companies to manage the activities better that can have a significant environmental impact.

In a clear analogy with the promotion of meta-standards for quality and environmental management systems, the process of fostering of international meta-standards for CSR, also called social accountability standards, began. Indeed, research on these meta-standards for CSR has become a major growth area in the field of business ethics (e.g. Gilbert and Rasche 2007; El Abboubi and Nicolopoulou 2012; Bres 2013). Nevertheless, the impact of SA8000 certification at a company level have not been systematically researched and consolidated yet (Grüninger 2009). This is clearly an under-researched field if the prominent literature on the adoption of other meta-standards such as ISO 9001 and ISO 14001 is considered (for recent reviews, see Heras-Saizarbitoria and Boiral 2013; Boiral et al. 2017a, b).

In order to try to fill this gap in the literature, the aim of the present chapter is to shed light on the adoption of the SA8000 meta-standards by Italian companies. Italy is the country with the higher intensity of SA800 certification. For that purpose, the remainder of this paper is arranged as follows. Following this introduction, a short review of the content and implications of the SA8000 standard for CSR is presented. In the third section the evolution of the international adoption of the SA8000 standard is analyzed. In the fourth section the case of the adoption of the SA8000 in Italy is introduced. In the fifth section the survey that was carried out is presented. In the sixth section the key findings are shown. Lastly, the conclusions of the survey are presented.

## 2 The SA8000 Standard

The SA8000 standard was first introduced in 1997 in the United States by the Council on Economic Priorities Accreditation Agency (CEPAA), with the aim of improving labour conditions on a global scale. Since 1998 the Social Accountability International ("SAI") organization, an affiliate of the Council on Economic Priorities ("CEP") has been responsible for the standard. The standard was updated in 2001 and 2008.

SA8000 is an auditable standard for third party verification, to ensure both ethical sourcing of products and goods and workplace conditions worldwide. It is a voluntary standard which can be implemented by all kinds of organizations, regardless of size, industry sector, geographic location or whether it is run for profit or as a non-profit. The standard seeks to guarantee working conditions and human rights in the workplace by providing basic requirements regarding the social responsibility of businesses. The nine requirements are the summarized in Table 1.

**Table 1** SA8000 requirements

| Social responsibility requirements of the SA8000 standard | |
|---|---|
| 1. Child labour | Employers must not hire children under the age of 15 years |
| 2. Forced labour | Employers cannot force workers to work against their will |
| 3. Health and safety | Employers must take protective measures to guarantee workers' health and safety |
| 4. Freedom of association and right to collective bargaining | Workers must have the freedom to bargain with employers (create and become members of trade union of their choice) |
| 5. Discrimination | Racial and other discrimination is forbidden |
| 6. Discipline | Employers must not use or support the use of disciplinary practices |
| 7. Compensation | Employees must be paid at least the minimum wage |
| 8. Working hours | The working time must be limited to 48 h per week and overtime at 12 h per week |
| 9. Management systems | The management system should be standardised |

*Source* Own elaboration based on www.sa-intl.org

SA8000 is primarily based on the various conventions and recommendations of the International Labour Organization (ILO) and on the Universal Declaration of Human Rights and the Convention on the Rights of the Child of the United Nations (UN).

According to its promoters, SA8000 is based on the ISO 9001 management system model. Although the structure of the standard is quite similar in many respects to the standards ISO 9000 and ISO 14001, some profound differences can be found. One major difference is that ISO management standards are process-driven standards for Quality and Environmental Management while the SA8000 standard refers, in some specific cases, to compliance with certain levels of performance or results in the business. In other words, it differs from ISO 9000 as it includes a set of performance requirements. For example, to earn certification to SA8000, employers have to pay sufficient wages to meet workers' basic needs, provide a safe working environment, avoid child or forced labour, and maintain a specific maximum amount of hours per week. Moreover, SA8000 audits involve not only ensuring that the processes are in place (as ISO 9000 and ISO14000 audits do) but also require extensive site visits and interviews with workers and other stakeholders to ensure compliance. Finally, it is important to stress that SA8000 implementation not only affects the company that chooses to adopt it but also affects suppliers and other firms in the supply chain of the company. Consequently, the principle of SA8000 creates a supply chain effect, not only being applied internally, but also being used as a tool to manage suppliers (Rohitratana 2002).

When implementing the standard organizations can choose between two different types of certificates; these refer to two different categories of implementation of the standard:

- *SA8000 Corporate Involvement Program ("CIP").* This category refers to companies which only buy and sell (particularly retailers, brand companies and wholesalers) and therefore have no manufacturing processes. These organizations have lower requirements and do not have external audits, but have to maintain a self-assessment report and are assisted by the SAI organization. The CIP is a two-level program that helps companies evaluate, implement and report their progress on the SA8000 standard. The two levels are:
  - *SA8000 Explorer.* The only requirement is for companies to evaluate their conformity with the standard as a CSR initiative through assessment audits.
  - *SA8000 Signatory.* The company has to promote and require its suppliers to implement the SA8000 standard, implement SA8000 over time throughout their supply chain and communicate their progress in implementation through SAI-verified public reports.
- *Certification to SA8000.* If the company is a manufacturer or a supplier it can be certified in compliance with SA8000 through audits conducted by SAI's accredited certification bodies.

A "guidance document" is issued by SAI, which provides instructions concerning the adaptation and basic compliance rules of SA8000. To foster implementation, SAI suggests dividing the process into four phases following Deming's (2000) Plan-Do-Check-Act model.

- *Plan*. In the first phase the organization chooses a certification firm—auditor—that is accredited by SAI, also known as a "certification body" and conducts an initial self-assessment.
- *Do*. During the second phase, any necessary improvements or modifications of production processes and management system are implemented by the company to meet the requirements of the standard. At this point, the auditor conducts a pre-audit and informs the company about any non-conformities which must be corrected.
- *Check*. Once the facility meets all the requirements, the actual audit takes place. In case of successful certification, the facility is awarded with the SA8000 standard for three years and can display the certificate.
- *Act*. The last phase is supposed to guarantee constant compliance with SA8000. Auditors make surveillance audits throughout the three year period and can withdraw certification. At the end of the three-year period, if the company wants to renew its certification, it must apply for an extension. To ensure dialogue during this period, employees can report their complaints in an anonymous way.

Some examples of certification agencies that are accredited by SAI are Det Norske Veritas, SGS-ICS, Bureau Veritas Quality International and AENOR. The cost of certification depends on the size, scope, and location of the organization and varies depending on the number of days and auditors needed to conduct the audit. The number of required audit days varies, but an average audit would consist of the following steps and related auditor days (Ciliberti et al.`2011)

- Document review: one or two days
- Office audit: one or two auditors for two–three days
- Witness audit: two auditors for two–three days
- Report writing: one–two days

This gives a total of 10–15 days per facility, while the costs typically range between \$500 and \$1500 per day (plus travel). There is also an application fee (depending on the number of employees) and an annual royalty. Over three years the total cost would range between \$20,000 and \$40,000 per company.

## 3 Evolution of the Adoption of the SA8000 Standard

In this brief section we will analyse how the SA8000 standard has evolved during the last decade and in which countries and type of organizations there has been more interest in this standard.

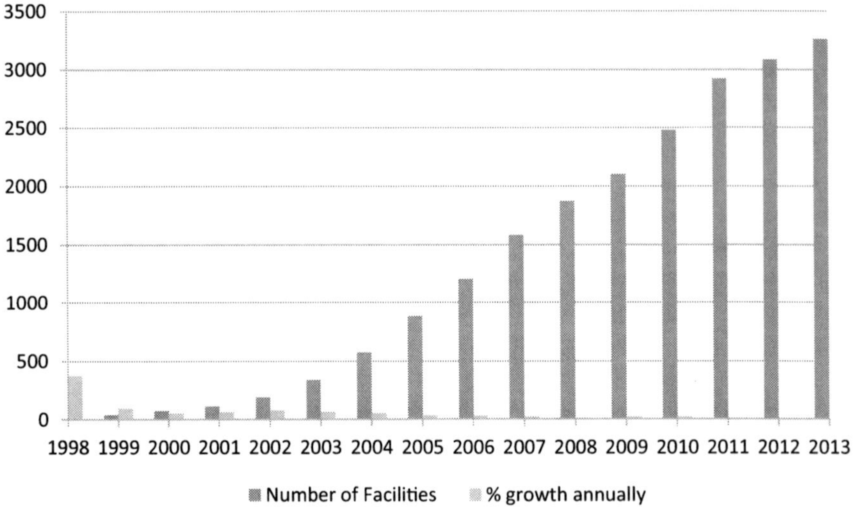


**Fig. 1** Evolution of the number of facilities certified by year. *Source* Own elaboration based on data from SAI. *Note* Updated 30th June 2013

According to SAI, by June 2013, 3258 certificates had been issued in 72 countries. As Fig. 1 shows, there was been an important growth in the number of certified facilities during the first decade (Fig. 2).

However, during the period 2008–2013 there seems to be a stabilization in the number of certificates. This fact can be linked, among other factors, to the economic crisis that has been and still is present in the European countries.

As previously mentioned, the implementation and maintenance costs of the SA8000 standard are high, so many companies decided not to apply for an extension when their certificate expired, because of the economic crisis. The decline in the growth of certified facilities may also be attributed to an end of a period of boom and the nature of the SA8000 standard.

More than 95% of the certified companies are located in Europe and Asia. 1555 certificates have been awarded in Europe and 1580 in Asia, representing 47% and 48% of all the certificates issued worldwide respectively. Although the organization responsible for the standard (SAIl) is based in the U.S., companies in that country have shown hardly any interest in the standard. In fact, there were only two certified companies in 2013.

Focusing on the Asian continent, Indian companies (24% of the certificates) and Chinese companies (17% of the certificates) predominate. Following these countries are Pakistan and Vietnam. The majority of these companies belong to the textile, garment and toy industries, all of which are labour intensive and where regulation of working conditions, and child and forced labour are relevant, aspects which are the focus of the SA8000 standard. In addition, in these countries there are also many certified facilities in the chemical sector.

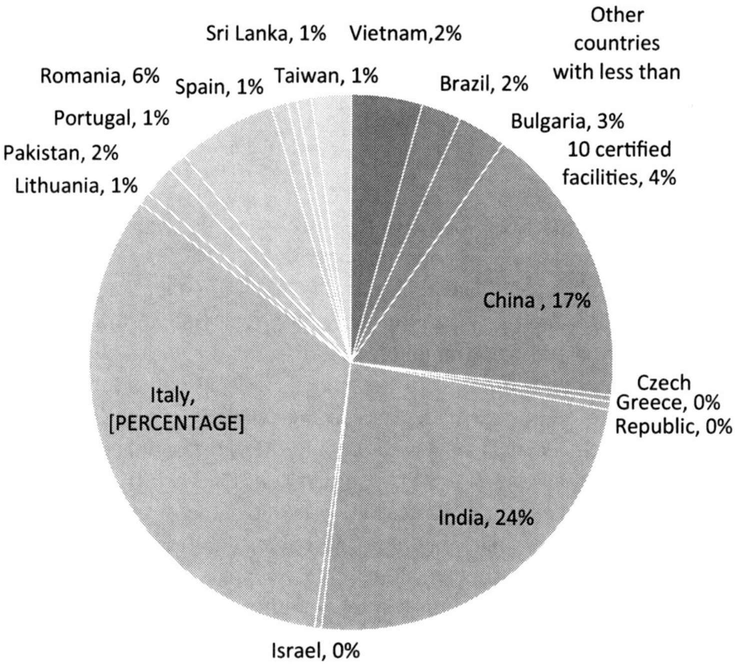


**Fig. 2** Certified facilities by country. *Source* Own elaboration based on data from SAI

In Europe, Italian certified organizations (1069) represent 33% of all the certificates. The reasons for this fact will be analysed in the following section. In contrast, the diffusion of the SA8000 standard in leading countries of the EU such as the UK, Germany and France has been minimal, (10 in UK, 9 in Germany and 4 in France). In Spain there were 33 certified organizations in 2013. It is important to highlight the policies promoted by the public authorities in the region of Aragon and the Provincial Council of Bizkaia.

Most of the companies with SA8000 certification are small and medium sized organizations; 67% of certified facilities have fewer than 250 employees. This is paradoxical, as SMEs have limited human and financial resources, which could be a barrier to their socially responsible behaviour. In addition, the costs of maintaining and implementing SA8000 are frequently too high for small and medium enterprises.

The reason for the predominance of SMEs is the fact that many small businesses deal with clients and employees at local level and depend on interpersonal relationships with stakeholders and their local involvement. Good reputation is a key to competitive advantage for SMEs and this is why these companies are certified SA8000 and behave in a socially responsible way. As Fassin (2009) remarks, SME managers are usually sensitive to activities that have an impact on their direct stakeholders, such as employees, customers and suppliers. SMEs often emphasize

**Table 2** General data

| | |
|---|---|
| Facilities currently certified | 3258 |
| Countries represented | 72 |
| Industries represented | 65 |
| Workers employed | 2,020,234 |

*Source* own elaboration based on data from SAI

responsible business practices, because they feel it is the right thing to do and responsible behaviour is seen as good management.

Finally, Table 2 shows the latest data available, as of June 2013, regarding the number of facilities which are currently certified, as well as the countries and industries represented and workers employed.

The number of organizations certified under SA8000 and similar codes continues to rise, while additional codes governing working conditions continue to emerge. Some academics, such as Hiscox et al. (2009), believe there is no rigorous evidence on whether adopting SA8000 or similar codes has any positive or negative impacts on staff turnover and absenteeism, product defect rates, sales growth or other measures of business performance. Moreover, they emphasize that a fierce debate is emerging about whether these codes represent substantive efforts to improve working conditions or simply symbolic efforts that allow organizations to score marketing points and counteract stakeholder pressure by merely filing some paperwork (Hiscox et al. 2009).

# 4 SA8000 Standard in Italy

## *4.1 National Context of CSR in Italy*

Social responsibility in Italian firms has roots dating back long before the emergence of the CSR movement during the 1990s. Several historical and structural reasons have characterized the Italian economic system and its relation to CSR, including the predominance of SMEs, the role of industrial districts and the importance of the cooperative movement (Tables 3, 4).

The Italian industrial system is characterized by a large majority of SMEs. A comparison with the average size of European companies in the industry, service and infrastructure sectors shows that the Italian system has an average of 4 employees per company compared to an average of 6 for the the 15 EU member states. In 2009 out of 4.5 million active firms, 95 per cent had fewer than 10 employees and overall employed 47 per cent of all workers (Istat,[1] 2011).

The activity of the Italian industrial system is geographically concentrated in industrial districts in Northern and Central Italy, along the Adriatic coast and in a

[1]The Italian National Institute of Statistics (ISTAT).

**Table 3** Certified facilities by size

| Workers employed | Number of facilities | % Total |
|---|---|---|
| 1–50 | 943 | 29 |
| 51–250 | 1183 | 36 |
| 251–1000 | 727 | 22 |
| >1000 | 400 | 12 |

*Source* own elaboration based on data from SAI

**Table 4** Study profile

| | |
|---|---|
| Population | 977 |
| Study sample | 645 |
| Replies obtained | 130 |
| Reply percentage | 20.16% |

*Source* own elaboration

few areas of the south. The northern part of Italy is more managerially orientated and influenced by European development. In some parts of Southern Italy there is still an important presence of illegal associations and this has a profound influence on trade and has prevented many businesses from developing. Northern Italy is where the major economic development has taken and is taking place. Research promoted by the Milan Chamber of Commerce shows that Northern Italy is more attentive to the environment and social issues than Southern Italy.

In this context, several initiatives on CSR have been promoted in response to increasing public attention to environmental protection, product safety and respect for human and worker rights (Perrini et al. 2006). These initiatives can be classified into three main categories:

- *Private Sector*: companies and other agents have taken several initiatives in CSR, which can be divided into four main groups.
    - *Reporting activities*: The introduction and spread of social reports in Italy has increased in recent years; many private and public institutions create a social report in order to provide evidence to the general public of their social commitment, values and initiatives.
    - *Adoption of management and certification standards*: CSR management systems such as ISO 14001, certification standards by third parties such as SA8000 and environmental and quality labels such as Eco-label.
    - *Financial investment*: One example of ethical finance is the first Ethical Bank, Banca Popolare Etica, founded in 1998, and in 2003 1.5 billion euro were managed through ethical funds (Ministry of Welfare, 2003).
    - *Other initiatives*: The creation of specific CSR awards, such as Oscar di Bilancio or Sodalitas Social Award, or the role of banks in promoting socially related initiatives.
- *Public Sector*: Regional and local governments have been deeply involved in promoting CSR through public policies. Most of these policies have been

focused on SMEs. Due to the large number of initiative, many of which are not relevant to this paper, only the most important initiatives related to CSR are mentioned. In June 2002, the Italian Ministry of Welfare, in collaboration with Bocconi University, carried out a project on Corporate Social Responsibility-Social Commitment (CSR-SC). The purpose of this project was to:

  - Promote CSR culture among companies.
  - Define a simple standard that firms can adopt on a voluntary basis in order to identify socially responsible behaviour.
  - Propose a list of relevant performance indicators to measure the social performance of companies.
  - Guarantee citizens that the reporting of corporate social commitment by companies is true and not misleading.

- *Corporative Association*: the Italian Union of Chambers of Commerce, (Unioncamere) has taken an active role in promoting CSR among companies working in Italy. An example of this participation is REBUS (RElation between BUsiness and Society), a project funded by the European Commission aimed at investigating the attention paid by SME managers to CSR.

## 4.2 *Adoption of the SA8000 Standard in Italy*

After this brief analysis of the CSR situation in Italy, we focus on the adoption of the SA8000 standard in this country. As stated, Italy is third in the ranking of SA8000 certifications and is the country in the world with a higher intensity of SA8000 certification in relative terms (if its GDP is taken into account). These figures are rather paradoxical, especially if it is taken into account that the SA8000 standard was designed for developing countries with labour intensive industries (Rohitratana 2002). The intense diffusion experienced by the SA8000 standard in Italian companies can be explained by to two main factors.

The first and most important is the strong support given by some regional governments to SA8000 certification. Many initiatives have been carried out, both at national and regional or local level, with the goal of promoting the culture of CSR and the adoption of specific practices and tools. Among the initiatives promoted at a regional level, the leading examples are:

- *Regional Act n. 20 of 2002*. This act was approved by the region of Umbria to establish a regional register of SA8000-certified companies and therefore promote a culture of CSR among companies. Inclusion in this register implied preferential treatment in financial support for training and certification as well as fiscal incentives.
- "*Lavoro Etico*". This initiative was launched by the Centre for Economic Development and Innovation (CISE) in the region of Emilia Romagna. The aim of this project was to create a social label linked to the SA8000 standard.

- "*Fabrica Etica*" project, implemented by the Tuscany Region. This project started in June 2000 with the aim of promoting SA8000 certifications among SMEs. The specific measures of promotion included training programs on CSR and SA8000, as well as financial support for companies which wanted to be certified SA8000, covering up to 50% of their consulting and certification costs. Moreover, in 2006 the Tuscan regional government developed specific legislation on CSR and established several incentive measures for companies that gain CSR certification, such as tax breaks or preferential treatment by the administration. These incentives have pushed many companies to apply for SA8000 certification.

It may seem strange that one of the richest and most developed regions of Italy, such as Tuscany, has dedicated so much effort to the promotion of a standard which is more appropriate to emerging countries. In order to understand this, we should refer to some facts that have had an impact in the media and that have pushed the regional authorities to develop promotion policies for SA8000.

As pointed out by Carey (2008) in 2001 the international media began to denounce cases of exploitation of working conditions of Chinese immigrants and child labour. Most of these cases referred to companies in the textile industry located in the Tuscany region; specifically in Prato, close to Florence, which is the largest community of Chinese immigrants in Italy and the third largest in the EU after Milan and Paris. A report published in 2004 by the Institute for Economic and Social Research (IRES) and the CGIL union,[2] presented some very disturbing information about the extent of the problem of child labour in businesses run by Italian immigrants. This evidence led the regional government to promote the adoption of the SA8000 standard with the purpose of improving working conditions and labour practices in Tuscan SMEs.

The second factor that explains the intense diffusion of the SA8000 standard in Italian companies, particularly the cooperative organizations, is associated with the driving force of a cooperative organization of great relevance in Italy: Coop Italia (Heras-Saizarbitoria and Landín 2013).

Coop Italia is the largest Italian retail chain, with a 17.1% market share in the grocery market. It is a national consortium that carries out purchasing, marketing, and quality control activities for regional cooperatives. It controls more than 3000 food and non-food suppliers throughout the world (Tencati and Zsolnai 2009).

In December 1998, Coop Italy was certified according to SA8000, becoming the first European company to obtain this certification. It first adopted SA8000 internally and subsequently involved all the suppliers of Coop-labelled products in the adoption of the standard. In 2009 there were 400 suppliers participating in this initiative all over the world. Coop Italia's ethical commitment has been recognized at international and national level. As an example it has won the 2001 Corporate Conscience Award given by Social Accountability International (SAI).

---

[2]The Italian General Confederation of Labour (CGIL) is a national trade union centre in Italy.

Tencati and Zsolnai (2009) point out that thanks to the application of the SA8000 certification considerable improvements were secured in the working conditions in companies inspected both in Italy and abroad as regards salaries, discrimination and occupational health and safety. According to Heras-Saizarbitoria and Landín (2013) this statement cannot be confirmed with the existing documentation. However, what is certain is that thanks to Coop Italia the number of Italian SA8000 certified companies has increased.

As a result of these two factors there has been an intense adoption of the SA8000 standard in Italian industry. Figure 3 shows the sectoral distribution of the certified facilities. These companies belong to diverse sectors, where the importance of services must be emphasized. 59% of the facilities operate in a service industry. This is striking because companies in sectors such as business services, social services and cleaning services were not the primary target of the SA8000 standard. This is due to the support received from public administration in the implementation of the standard.

In other sectors where certification has been high, such as the transport sector and food services, the high number of certified facilities (5 and 11% respectively) seems to be influenced by the initiatives of Coop Italia. Finally, according to data from SAI, there is a high level of participation among cooperatives in the fields of social and environmental services (8 and 5% respectively).

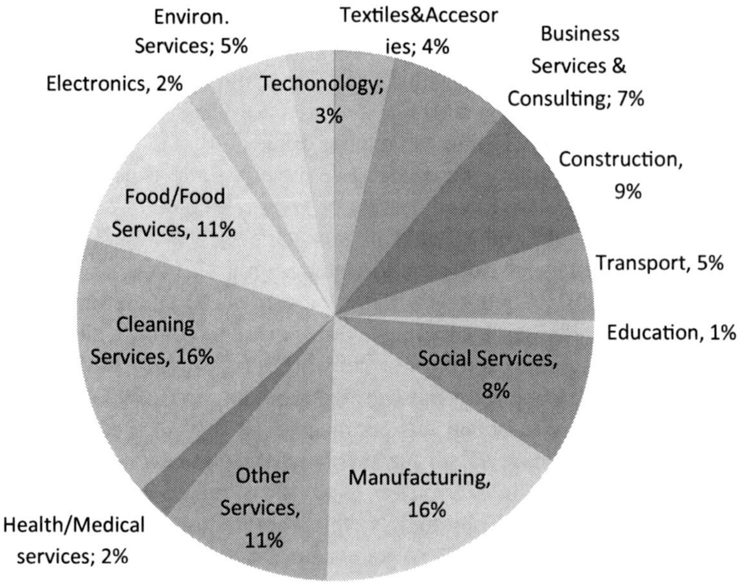


**Fig. 3** Italian certified facilities by industry. *Source* own elaboration based on data from SAI

## 5 The Adoption of the SA8000 Standard in Italian Companies: A Survey

The next part of this paper provides an overview of the adoption of the SA8000 standard in Italian companies through the analysis of the preliminary results of an empirical survey. The primary objective of the survey was to gather descriptive data and information on the adoption of the SA8000 standard in Italian organizations. More specifically, the work aimed to focus on the following topics:

1. The attention paid by companies to CSR issues
2. The motivational factors to implement SA8000
3. Difficulties and obstacles in implementing SA8000
4. Perceived advantages of SA8000

The approach for this investigation was to select a sample of 645 facilities. The data relating to 30 September 2012 was obtained from the database of SAAS—Social Accountability Accreditation Service. The first step was to gather as much information as possible about the facilities (web site, telephone, fax, person in charge of the SA8000 standard) in order to contact them.

To conduct the survey, a structured questionnaire was prepared based on the previous literature of the field and on previous work by the authors. The questionnaire was kept as short and clear as possible, with particular attention paid to the wording in order to ensure the maximum response rate. The survey was carried out between March and May 2014. Organizations were first contacted by e-mail with a direct link to complete the questionnaire online. A month later an e-mail reminder was sent. In a second step, the organizations were contacted by phone, and, if willing to participate in the survey, the questionnaire was sent to the appropriate person by email. Finally, a last reminder letter was sent.

Overall, 977 organizations were considered but of these the e-mail contact was found of only 645. The number of returned questionnaires from the 645 organizations was 130, representing a return rate of approximately 13.3% of the population and 20.16% of the sample. This rate of response enables us to draw statistically significant conclusions about SA8000 in Italian organizations. In this work in progress, all the information collected was analyzed using descriptive statistics.

The main characteristics of the companies that took part in the study are described below.

Most of the companies that are certified as compliant with SA8000 are SMEs (70%), as shown in Fig. 4. Organizations with fewer than 10 workers account for 15% of the organizations, and the same percentage are large enterprises.

The results in Fig. 5 show that the prevailing market of SA8000 certified facilities is national (51% of the facilities). However 21% have a local dimension and 18% operate at the international level. This circumstance can be explained by

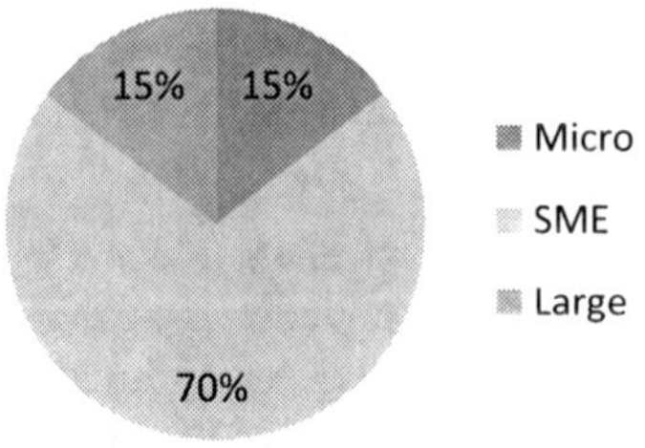


**Fig. 4** Number of employees of the responding organizations. *Source* own elaboration from data collected by the survey

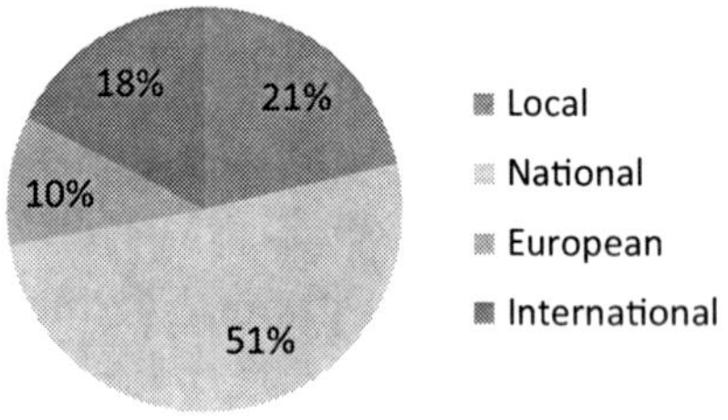


**Fig. 5** Internationalization level of the responding organizations. *Source* own elaboration from data collected by the survey

the relatively small size of the organizations participating in our survey, as well as the public policies promoted in the country in order to incentivize certification.

## 6 Findings of the Survey

After presenting the main characteristics of the organizations that took part in the survey, we now examine the responses related to general aspects of CSR. First of all, respondents were asked to indicate what CSR meant to them. A list of several issues commonly associated with CSR was proposed and each item on the list invited an evaluation of its significance on a scale ranging from 1 to 5.

According to data collected from the 130 surveys, the average response to the attention paid to CSR issues by organizations was 3.88 on a scale from 1 to 5, which can be considered high. Facilities gave almost the same importance to the aspects illustrated in Fig. 6. However, organizations gave slightly more importance to the fact that business should have a social responsibility beyond making a profit and pay less attention to the employee morale as the first priority. Even if the difference is not striking, this result is surprising since SA8000 focuses specifically on the working conditions and rights of the organization's employees.

Figure 7 indicates the main factors that have pushed organizations to implement the SA8000 standard. The items that received higher average scores are the top managers' social responsibility and ethical concerns and public demonstration of CSR commitment. Therefore, SA8000 certification is seen as a strategic tool that can positively affect the organizations by improving their corporate image and brand reputation (Fig. 8).

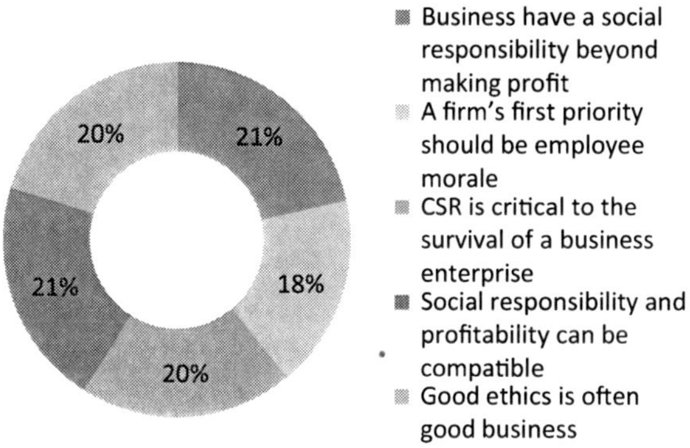


**Fig. 6** The importance of CSR issues for Italian certified organizations. *Source* Own elaboration from data collected by the survey

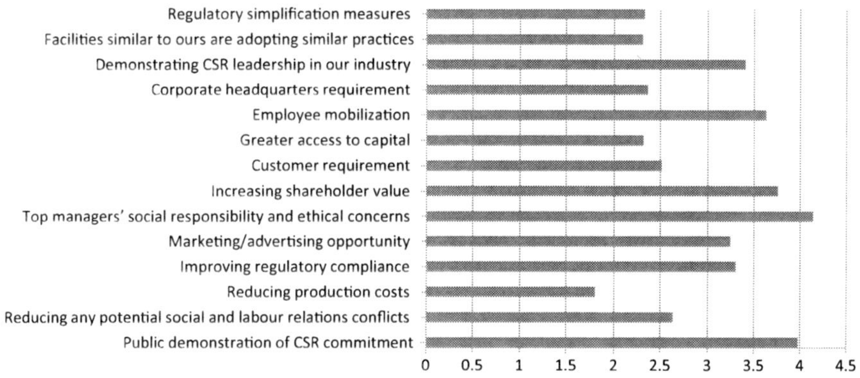


**Fig. 7** Main drivers for SA8000. *Source* Own elaboration from data collected by the survey

On the hand, the least important item is the reduction of production costs. This result is not striking since cost saving is a driver that is more appropriate for quality management standards such as ISO 9001. Furthermore, SA8000 is not perceived as a very useful tool to attract financial resources. This is linked to the fact that socially responsible investment is almost unknown and still has a long way to go. According to specialists, socially responsible investment funds appeared in the Italian market in the 90s but the sector has never flourished, and is currently worth less than 2 billion euro.

When considering the size of responding organizations, no substantial differences emerge in the drivers for the implementation of the SA8000 standard, with the exception of micro organizations, which are more influenced than SMEs and large corporations by the public demonstration of CSR commitment. This can be explained by the fact that small enterprises use the standard as a tool to improve

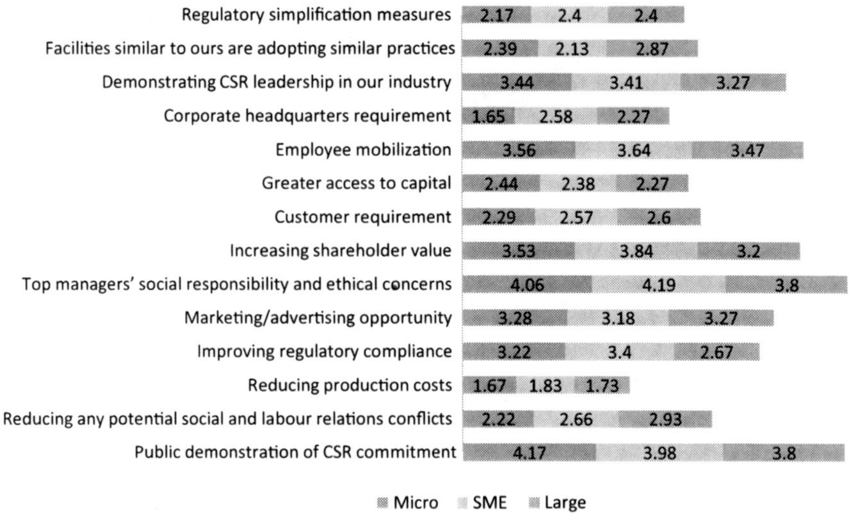


**Fig. 8** Main drivers for SA8000 by size of responding organizations. *Source* Own elaboration from data collected by the survey

their public image, build confidence and strengthen relations with stakeholders at the local level (mainly customers and suppliers).

SMEs are the organizations most concerned to adopt the SA8000 standard because of managers' concerns and in order to increase shareholder value.

Finally, large firms attach more importance than micro and SMEs to the reduction of potential labour and social conflicts. This is because multinationals that outsource their production to countries with cheaper labour costs, where exploitation and poor working conditions are rife, face a greater risk of labour disputes. These organizations are frequently the focus of social organizations, and therefore the SA8000 standard can be an interesting tool to reduce social conflicts.

The integration of standards obviously comes up against difficulties when putting them into practice. In order to gain a better understanding of the real causes of the problems in the integration process, organizations were asked to rank to the extent to which the problems shown in Fig. 9 had affected the implementation of the SA8000 standard, on a scale from 1 to 5.

Overall, respondents do not appear to be significantly affected by the proposed difficulties, since all of the factors analysed except one are below the average of 3. The most outstanding issue is the difficulty to implement measures along the supply chain.

The principles of SA8000 tend to create a supply chain effect, as it is used as a tool to manage suppliers. A company seeking certification must implement proper steps to ensure that its suppliers comply with the standard. However, not every firm is in a position to persuade supply chain partners to implement SA8000, as some (especially SMEs) lack the power to influence or force partners.

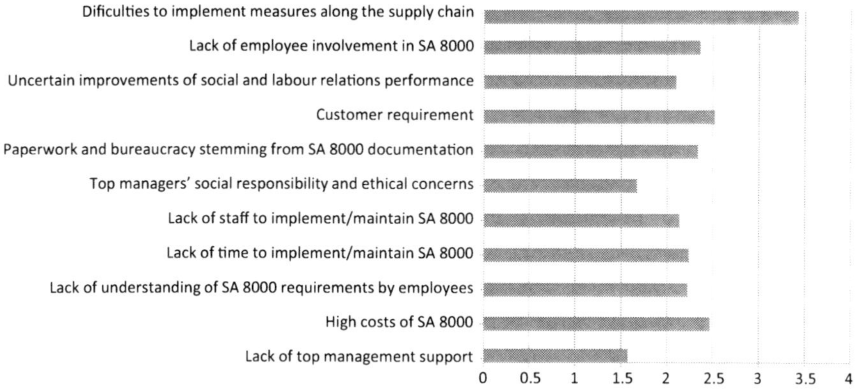


**Fig. 9** Obstacles and problems associated to SA8000. *Source* Own elaboration from data collected by the survey

Moreover, from a supplier's perspective, Stigzelius and Mark-Herbert (2009) identify the lack of support from buyers as an obstacle to the implementation of SA8000. Buyers do not share the costs incurred by suppliers to become compliant with the standard. In addition, there are no contracts to ensure that complying with standards represents a safe investment. Combining the standard's requirements with attractive prices for buyers is therefore difficult. This may be a reason for the lack of involvement on the supplier side, making it difficult to implement measures along the supply chain.

Secondly, customer requirement and the high costs related to SA8000 are difficulties encountered in the integration process. In the telephone calls about the survey, all of the responding organizations which stated that they had not renewed the certification said that the reason was due to the high costs that SA8000 involved, and which they were not able to cope with during these difficult years of economic crisis.

According to the results shown in Fig. 10, smaller organizations appear to have more significant economic difficulties with their certification than larger firms. For smaller companies that do not have the money and buying power of the large multinationals, the implementation of the SA8000 standard is especially expensive and time-consuming. On the other hand, for the other items, large firms give a higher score than micro and SMEs.

The final section of this study focuses on the benefits of SA8000 certification, both at a general level and more specifically by size of organization.

The results shown in Fig. 11 indicate the most significant benefit of the implementation of SA8000 standard is the greater commitment to social responsibility and ethics. This element is very similar to the most important certification driver, so that we can conclude that SA8000 is an effective tool. Improved employee training and improved social responsibility management practices are the other items that receive a higher than average score from the respondents (Fig. 12).

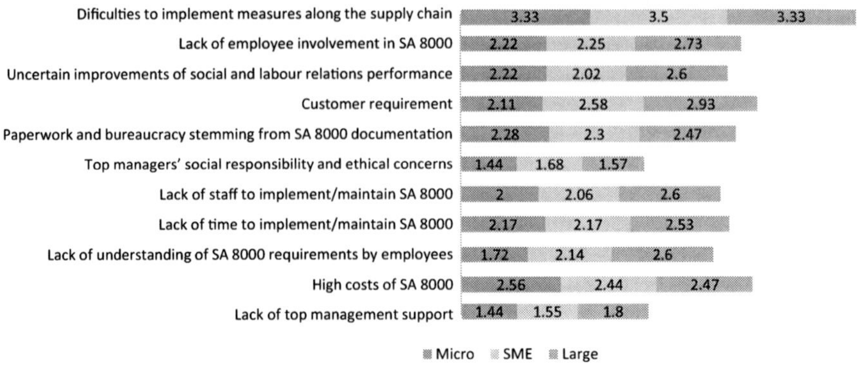


**Fig. 10** Obstacles and problems associated to SA8000 by size of responding organizations. *Source* Own elaboration from data collected by the survey

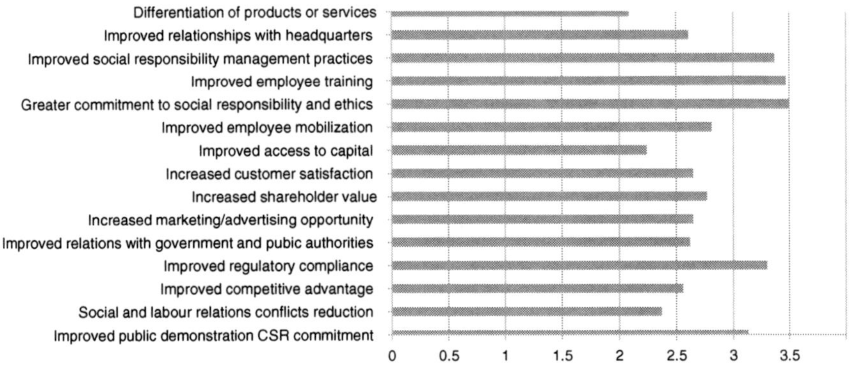


**Fig. 11** Benefits from the implementation of SA8000. *Source* Own elaboration from data collected by the survey

Conversely, it would appear that SA8000 is not very effective at ensuring easier access to capital or as a strategic tool for differentiation of products and services. This reaffirms the previously mentioned fact that socially responsible investment is still an under-developed area of the Italian economy.

Moreover, organizations do not consider SA8000 a marketing tool for differentiation, since the purpose of the standard is to enforce the image and reputation of the brand or company and to improve the labour conditions within it rather than an instrument to distinguish a product or service and therefore increase sales or profit.

When considering the size of the responding company, small and medium organizations seemed to perceive slightly greater benefits from the implementation of SA8000. The average scores for all benefits are 2.61 and 2.64 for large and micro organizations respectively, while for SMEs this figure is 2.91.

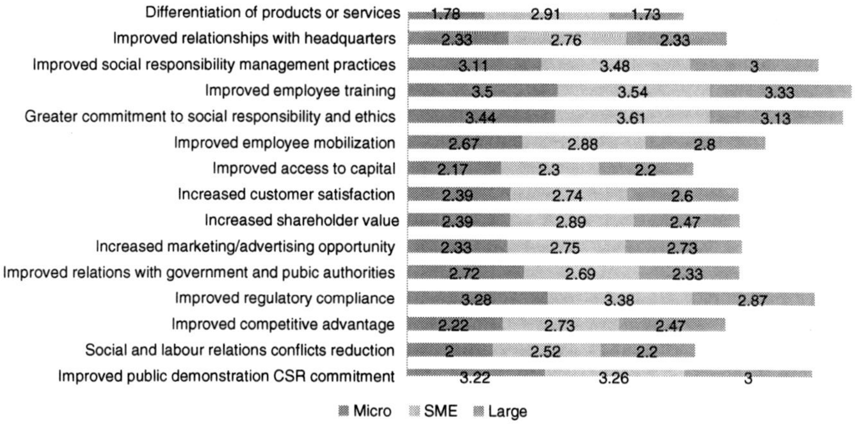


**Fig. 12** Benefits from the implementation of SA8000 by size of responding organizations. *Source* Own elaboration from data collected by the survey

# 7 Conclusions

This research aimed to shed light on the paradoxical case of the adoption of SA8000 in Italy as an example of the complex issue of the adoption of this CSR meta-standards. The active role of Italian public administrations in the promotion of this type of meta-standard is underlined. Regarding the data obtained in the survey, in which a relevant number of Italian SA8000 certified companies took part, which suggests the results are reliable, the following specific conclusions can be drawn:

- Responding organizations are mainly small and medium sized. The average size of respondents' firms, 178 employees, is larger than Italian firms overall, probably because CSR costs, the paperwork related to SA8000 and the lack of time to implement this system tend to discourage microenterprises.
- CSR awareness and understanding among responding organizations seems to be reasonably good. Organizations believe business should have a social responsibility beyond making a profit but also find social responsibility and profitability can be compatible.
- SA8000 certification drivers are both strategy and sustainability-related while financial drivers score poorly. The principle reasons which make companies adopt the standard are top managers' social responsibility and ethical concerns and public demonstration of CSR commitment.
- Respondents did not encounter many relevant difficulties during the certification process. The most outstanding issues were the difficulty in implementing measures along the supply chain and the high costs related to SA8000. Small organizations had more difficulties when adopting the standard, since the high costs of certification and maintenance often present a barrier for them, and smaller organizations tend to have priorities other than CSR.

- Finally, regarding the supposed benefits of SA8000 certification, the surveyed organizations emphasize greater commitment to social responsibility and ethics, improved employee training and improved social responsibility management practices. It is also worth mentioning that SMEs perceive more benefits from the implementation of SA8000 than large and micro facilities.

All in all, the widespread diffusion of the SA8000 standard in Italy has much to do with the strong certification culture of that country. This has also emerged from our survey, given that many organizations have more than just SA8000 certification. In fact, 87% of the respondents also had another certification, the most common of which were ISO 9001 and ISO 14001, the global meta-standards for quality management and environmental management, respectively. SA8000 certification is supposed to send a positive signal to markets on the adoption of CSR practices and tends to reduce the reputational risk. Nevertheless, as underlined by Boiral et al. (2017b) stakeholders are rarely aware of the real effect of a CSR certification on the internalization of CSR practices. In many certified organisations, it seems that SA8000 certification is primarily used as a commercial and legitimization tool rather than as a management system aimed at internalizing CSR issues into dailies activities. This type of symbolic rather than substantial use of certifiable management systems is quite common in ISO certifications such as ISO 14001 or ISO 9001 (Christmann and Taylor 2006; Heras-Saizarbitoria and Boiral 2015; Heras-Saizarbitoria et al. 2013), and the same remark can certainly applies to SA8000 (Boiral et al. 2017b). This observation is in line with the neo-institutional approach of certifiable management systems. This approach emphasises the frequent disconnection between the search for social legitimacy through recognized standards such as SA8000 or ISO 14001 and the internal practices which are not necessarily in line with those standards (Heras-Saizarbitoria and Boiral 2013; Boiral et al. 2017b). As a result, managers and employees of certified organizations tend to pay lip services to the standards they are supposed to implement.

This work has some managerial implications. Managers concerned by the successful adoption of the SA8000 standard should clarify the reasons and the institutional pressures underlying their adoption. They should probably not consider SA8000 as a mere commercial certificate with market and reputational benefits, considering the recent concerns raised by relevant stakeholders. Overall, the internal motivation to adopt certifiable standards is correlated with the positive outcomes of these systems (Heras-Saizarbitoria and Boiral 2015; Boiral et al. 2017a). As underlined by Boiral et al. (2017b) government and standardization agencies should also develop more stringent training programs and accreditation procedures for the auditors involved in SA8000 certification and a more in-depth analysis of the real implications of the public incentives aimed at fostering the adoption of this type of CSR standard.

The limitations of this study provide opportunities for future research. Like most of the quantitative studies focused on the study of the adoption of ISO standards, the information used in this paper is based on the perceptions of managers who have taken part in the process of adoption of the SA8000 standard and therefore the results may be influenced by the social desirability bias of the respondents, a bias related to the personal interests of the respondents in underlining the success of the adoption of the CSR standard. Therefore, as emphasized by Heras-Saizarbitoria et al. (2015) for the case of environmental management standards, more research should be carried out with companies with SA8000 certification based on information provided by other stakeholders such as employees. Interviews conducted outside the working environment can also help to reduce the organizational and institutional pressures underlying the social desirability bias in research on certifiable standards (Boiral 2003).

Although SA8000 has been considered to be one of the best CSR initiatives in the scholarly literature (Gilbert and Rasche 2007), some concerns regarding this type of CSR standard should be also mentioned. From the very beginning of the dissemination of this standard there has been a deep concern among stakeholders—including NGOs (e.g. LARIC 1999), some practitioners (e.g. Zuckerman 1998) and even by the media (e.g. Chan Chan 2005)—about the real potential of this type of standard. As documented by Boiral et al. (2016), recent incidents in CSR certified factories, such as the tragedies at Rana Plaza in Bangladesh in 2013 (which had Fair Trade USA certification)—with the death of 1127 workers and 2437 injured—and at Ali Enterprises in Pakistan—SA8000 certified three weeks before 258 workers were killed in a fire—have deepened the concerns of those stakeholders about CSR standards such as SA8000 and the failure of third party certification. Taking into account the lack of research regarding the adoption process of the SA8000 standard, future research could investigate the perceptions of SA8000 among different stakeholders, their expectations of this CSR standard and whether SA8000 is really integrated into the daily activities of organizations or whether it could be related to a sort of CSR-washing practice. The real effect of the CSR standard in the improvement of the social performance of certified companies from a rigorous and independent perspective should be also researched.

**Acknowledgements** This work has been carried out for the Research Group funded by the Basque Government (Grupos de Investigación del Sistema Universitario Vasco; IT1073-16).

## References

Boiral O, Heras-Saizarbitoria I (2015) The SAGE encyclopedia of quality and the service economy. Sage, Thousand Oaks

Boiral O (2003) ISO 9000: outside the iron cage. Organ Sci 14(6):720–737

Boiral O, Guillaumie L, Heras-Saizarbitoria I, Tayo Tene CV (2017a) Adoption and outcomes of ISO 14001: a systematic review. Int J Manage Rev (in press)

Boiral O, Heras-Saizarbitoria I, Testa F (2017b) SA8000 as CSR-washing? The role of stakeholder pressures. Corp Soc Responsib Environ Manage 24(1):57–70

Bres LP (2013, January) Membership strategies in pluralistic organizations: the case of ISO 26000. In: Academy of management proceedings (vol 2013, no. 1). Academy of Management, p 15343

Carey C (2008) Governmental use of voluntary standards case study 10: Tuscany Region (Italy) and the SA8000. Standard for Social Accountability, London
Chan A (2005) Recent trends in Chinese labour issues—signs of change. China Perspect 57:23–31
Christmann P, Taylor G (2006) Firm self-regulation through international certifiable standards: determinants of symbolic versus substantive implementation. J Int Bus Stud 37(6):863–878
Ciliberti F, De Haan J, De Groot G, Pontrandolfo P (2011) CSR codes and the principal-agent problem in supply chains: four case studies. J Clean Prod 19(8):885–894
El Abboubi M, Nicolopoulou K (2012). International social-related accountability standards: using ANT towards a multi-stakeholder analysis. M@n@gement 15(4):392–414
Fassin Y (2009) The stakeholder model refined. J Bus Ethics 84(1):113–135
Gilbert DU, Rasche A (2007) Discourse ethics and social accountability: the ethics of SA8000. Bus Ethics Q 187–216
Grüninger B (2009) The impact of SA8000 in Brazil. In: SA8000: the first decade: implementation, influence, and impact. Greenleaf Publishing in Association with GSE Research, vol 172(180), pp 172–180
Hahn R (2013) ISO 26000 and the standardization of strategic management processes for sustainability and corporate social responsibility. Bus Strategy Environ 22:442–455
Heras-Saizarbitoria I, Boiral O (2013) ISO 9001 and ISO 14001: towards a research agenda on management system standards. Int J Manage Rev 15(1):47–65
Heras-Saizarbitoria I, Boiral O (2015) Symbolic adoption of ISO 9000 in small and medium-sized enterprises: the role of internal contingencies. Int Small Bus J 33(3):299–320
Heras-Saizarbitoria I, Arana G, Boiral O (2015) Outcomes of environmental management systems: the role of motivations and firms' characteristics. Bus Strategy Environ (in press)
Heras-Saizarbitoria I, Dogui K, Boiral O (2013) Shedding light on ISO 14001 certification audits. J Clean Prod 51:88–98
Heras-Saizarbitoria IH, Landín GA (2013) La responsabilidad social corporativa y la norma SA8000: un análisis de su adopción en las organizaciones cooperativas. Revista vasca de economía social = Gizarte ekonomiaren euskal aldizkaria (9):31–52
Hiscox MJ, Schwartz C, Toffel MW (2009) Evaluating the impact of SA8000 certification. SA8000: First Decade: Implement Influ Impact 1(78):147–165
LARIC (Labour Rights in China) (1999) No illusions: against the global cosmetic SA8000. Asia Monitor Resource Center, Hong Kong
Perrini F, Pogutz S, Tencati A (2006) Corporate social responsibility in Italy: state of the art. J Bus Strat 23(1):65–91
Rohitratana K (2002) SA8000: a tool to improve quality of life. Manage Audit J 17(1/2):60–64
Stigzelius I, Mark-Herbert C (2009) Tailoring corporate responsibility to suppliers: managing SA8000 in Indian garment manufacturing. Scand J Manage 25(1):46–56
Tencati AY, Zsolnai L (2009) The collaborative enterprise. J Bus Ethics 85:367–376
Zuckerman A (1998) Do we really need a social accountability standard. Purchasing 125:24–27

# Adopting ISO/TS 16949 and IATF 16949 Standards: An Exploratory and Preliminary Study

**Iker Laskurain, German Arana and Iñaki Heras-Saizarbitoria**

**Abstract** ISO/TS 16949 was launched in 1999 by the International Automotive Task Force (IATF) with the aim of harmonizing the diverse assessment and certification systems worldwide in the supply chain for the automotive sector. ISO/TS 16949 aimed to eliminate the need of the automotive suppliers for multiple sectoral certifications such as QS9000, VDA6 and EAQF. The fourth version of the standard—published in October 2016—has been renamed IATF 16949. Despite the relative success of this standard for a very relevant industry such as the automotive sector, in the scholarly literature the adoption of the ISO/TS 16949 standard has been under-researched. In order to fill this gap, the aim of this Chapter is to analyze the critical points of working with a management system based on the ISO/TS 16964 or the IATF 16949 from a scholarly perspective, considering the strengths and flaws of the adoption of the standard. For that purpose, a qualitative exploratory study was designed—which is still in progress because of the transitionary nature of the standards. The research focuses on four case studies, one Original Equipment Manufacturer of the automotive sector and one manufacturer from each of Tier 1, Tier 2 and Tier 3 of the automotive sector. The perspectives of other main stakeholders, such as auditors and consultants directly involved in the implementation and certification of ISO/TS 16949 and IATF 16949 were also considered.

**Keywords** ISO/TS 16949 · IATF · ISO 9001 · Metasystem · Quality management · Automotive sector

I. Laskurain (✉) · G. Arana · I. Heras-Saizarbitoria
Department of Management, University of the Basque Country
UPV-EHU, San Sebastián, Spain
e-mail: iker.laskurain@ehu.es

G. Arana
e-mail: g.arana@ehu.es

I. Heras-Saizarbitoria
e-mail: iheras@ehu.es

G. Arana
E.U. Ingeniería de Gipuzkoa, University of the Basque Country
UPV-EHU, Plaza Europa 1, 20018 San Sebastián, Spain

 
I. Heras-Saizarbitoria (ed.), *ISO 9001, ISO 14001, and New Management Standards*, Measuring Operations Performance, https://doi.org/10.1007/978-3-319-65675-5_8

# 1 Introduction

Standardization is a mechanism of coordination and an instrument of regulation, comparable to other instruments, such as public regulations, markets and hierarchies or formal organizations (Antonelli 1994; Brunsson and Jacobsson 2000; Heras-Saizarbitoria and Boiral 2013). In management activities, the concept of a management standard, management system standard or meta-standard has been defined as a list of design rules to guide the creation of entire classes of management systems. Since systems theorists use the term metasystem for lists of this type, it follows that this type of management standard should be referred to as a meta-standard (Uzumeri 1997, p. 22).

These meta-standards relate to a very wide range of business activity, such as quality management, environmental management, the prevention of occupational hazards, the provision of health and safety regulations in the workplace and corporate social responsibility (Heras-Saizarbitoria and Boiral 2013). Among quality management standards, ISO 9001 was designed to be adopted in any sector and has been widely disseminated. By the end of 2015, there were 1,033,936 firms with this certificate in the world (ISO 2016).

However, in some sectors there are specific requirements. For this reason, the International Automotive Task Force (IATF) and the ISO created ISO/TS 16964:1999, trying to eliminate the need for automotive suppliers to seek multiple sector certifications, such as QS9000, VDA6, EAQF and AVSQ (Kartha 2004). Adopting this international technical specification is important in the automotive sector because companies that supply products to automotive manufacturers have to show that they have reached the official standard (Lupo 2002). This ISO technical specification, jointly developed, serves as a common automotive quality system requirement. It specifies the quality system requirements for the design, development, production, installation and servicing of automotive-related products (Kartha 2004).

The dissemination of this standard among automotive suppliers has been very important and it is still in a growth phase, reaching 62,944 certificates in 2015, as can be seen in Fig. 1. Nevertheless, in the scholarly literature the adoption of the ISO/TS 16949 standard has been under-researched, except for a few papers on the adoption of relevant standard (Kartha 2004; Ostadi et al. 2010; Singh 2014). In order to fill this gap, the aim of this Chapter is to analyze the critical points of working with a management system based on ISO/TS 16964. For this purpose, a qualitative exploratory study was designed—which is still on-going, because of the transitory stage of the standard.

The rest of the chapter is organized as follows. In the next section the ISO/TS 16949 standard is described. The literature review is summarized in the third section. The research methodology is summarized in the fourth section. In the fifth section, the analysis and discussion of the work in progress are included. Finally, in the sixth section the preliminary conclusions about stakeholders interested in this adoption and certification process are presented.

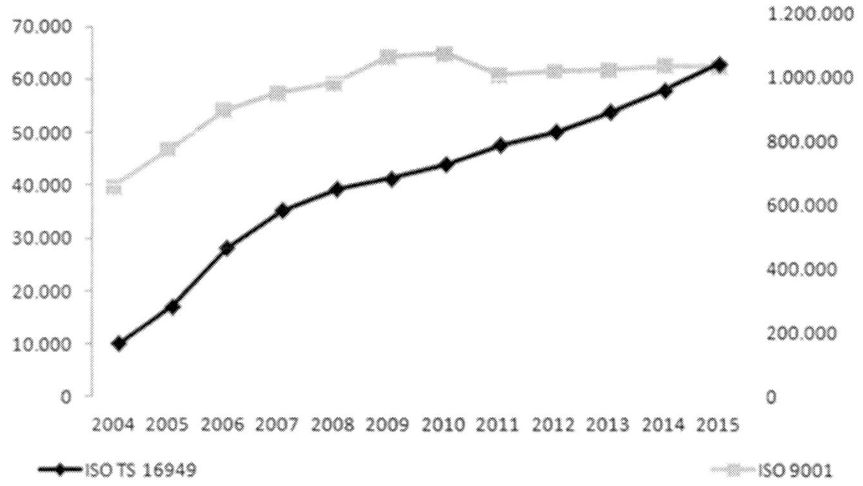


**Fig. 1** Dissemination of the ISO/TS 16964 and the ISO 9001. *Source* Elaborated by the authors with data obtained from ISO (2016) and Hys (2015)

## 2 ISO/TS 16949:2009 and the New IATF 16949:2016

ISO/TS 16949 was originally created in 1999 by the IATF with the aim of harmonizing the different assessment and certification systems worldwide in the supply chain for the automotive sector. ISO/TS 16949 introduced a common set of techniques and methods for common product and process development for automotive manufacturing, trying to eliminate the need of the automotive suppliers for multiple sector certifications, such as QS9000, VDA6, EAQF and AVSQ (Kartha 2004; IATF 2016). This was a very important field of work of the IATF members BMW Group, FCA US LLC, Daimler AG, FCA Italy Spa, Ford Motor Company, General Motors Company, PSA Group, Renault, Volkswagen AG and the vehicle manufacturers' trade associations AIAG, ANFIA, FIEV, SMMT and VDA. ISO/TS 16949 is considered in the literature to be a more demanding management standard than ISO 9001 as it contains specific requirements for the sector (Vaxevanidis et al. 2006; Singh 2014). Nevertheless, this issue has not been studies from a scholarly perspective using empirical perspective.

The fourth and latest version of the standard, IATF 16949:2016, was published in October 2016 and replaces the ISO/TS 16949:2009. The new standard is based on the new high level structure to maintain consistency, align different management system standards, offer matching sub-clauses against the top level structure and apply common language across all standards. The technical specification of IATF 16949:2016 is synchronised with the ISO 9001:2015 standard. It establishes detailed requirements concerning the quality management system during the production process as well as assembly and maintenance of products connected with the automotive industry. This revision also incorporates requirements by IATF for

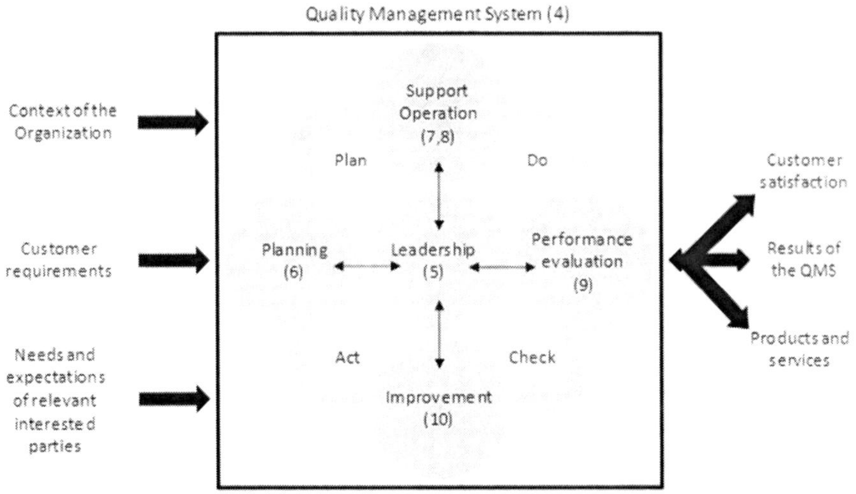


**Fig. 2** The PDCA version in the IATF 16949:2016. *Source* Elaborated by the authors based on BSI (2017)

original manufacturers (OEM) and customer-specific requirements (CSR) among other requirements (see Fig. 2). The chapters are similar to those of ISO 9001:2015 but include some relevant definitions for the automotive industry. At present, almost all companies are moving to the new version of the standard.

As shown in Fig. 2, the standard contains supplementary requirements specific to the automotive industry rather than being a stand-alone QMS (BSI 2017). For its promoters, the goal of this standard is the development of a global management system that provides for continual improvement, emphasizing defect prevention and the reduction of variation and waste in the supply chain. It emphasizes a process approach, and commitment to quality on the part of top management.

## 3 Theoretical Framework and Research Question

Laosirihongthong and Dangayach (2005), using a survey carried out in companies from India and Thailand, pointed out that companies from automotive sector must develop and maintain a high degree of coherence among competitive priorities, order-winning criteria, and improvement activities, due to the competitive market and customer pressures. In this context, the companies of this sector need to meet the requirements for standardisation of their customers (Kartha 2004).

In relation to the coordination between different tiers in the manufacturing process, some authors explain that the ISO TS 16949 standard eliminates redundancy, cost and administrative burdens imposed by multiple standard requirements (Ostadi et al. 2010). These aspects and the customer requirements are highlighted in the conclusion of a study carried out in the Association of South East Asian Nations, to explain why almost all companies that implement Total Quality Management (TQM) in the automotive sector had implemented ISO 9001:2000 and ISO/TS 16949 (Punnakitikashem et al. 2010). However, Hys (2014) found out that only 27.69% of Polish automotive companies have adopted ISO/TS 16949, VDA, AVSQ, QS 9000 or EAQF. Special requirements create the need to analyze the implications for management of certified automotive companies, but there are not many studies analyzing this aspect. Bevilacqua et al. (2011) analyzed the adoption of ISO/TS 16949:2009 in the production of stainless steel tubes for the automotive sector in an in-depth case study. The study, carried out on a floor-shop level and very applied perspective, showed that he application of ISO/TS 16949 led to a number of corrective actions, such us increasing controls on blade quality.

Regarding the adoption process, Cauchick Miguel et al. (2011) published a case study about the implementation of ISO/TS 16949 in a multinational company. The authors explained that the company moved from ISO 9000 to ISO/TS 16949. ISO 9001 and its adopted strategy facilitated the adoption of ISO/TS 16949, identifying the necessary work, allocating resources to conduct the work, and planning the auditing in order to achieve implementation goals. Ostadi et al. (2010) stated that the standard is focused on the uniqueness of each automotive supplier's process, while providing critical tools to help the companies to meet better customer-specific requirements. Hoyle (2005) added that it also emphasized defect prevention, and reducing variation and waste in the supply chain. For these reasons, in the literature for the automotive sector, ISO/TS 16949 is considered to be a more demanding management standard than ISO 9001 (Vaxevanidis et al. 2006; Singh 2014). Nevertheless, as has been underlined in this literature review, there is a clear gap in the scholarly literature of both qualitative and quantitative works that shed light on the real effect of adopting ISO/TS 16949 or the new IATF 16949. As it has been shown, the scattered works identified in the literature were carried out either from a practitioner perspective or from a very limited scholarly perspective. Those previous studies do not focus on the flaws and weaknesses of the adoption of the meta-standards. The aim of this research in progress is to analyze the critical points of working with a management system based on the ISO/TS 16964 or the IATF 16949 from a scholarly perspective, considering the strengths and flaws in the adoption of the standard.

## 4 Empirical Method and Field Work

Considering the objectives of the study, an exploratory qualitative study was designed based on case study methodology. This methodology was selected owing to its suitability when analyzing the complex process of ISO/TS 16949 adoption, in which—as has already been stated—diverse agents and actors interact. Research of a descriptive nature was planned, albeit mainly exploratory, that facilitates greater penetration and understanding of the subject, so as to try and identify propositions that are might be generalized in terms of the practices observed (Eisenhardt 1989; Yin 2003).

Four case studies of an Original Equipment Manufacturer (OEM), and one each of a Tier 1, a Tier 2 and a Tier 3 manufacturer, were carried out in total in manufacturing companies either ISO/TS 16949 implemented and certified or with companies in the process of adoption of the IATF 16949 but without certification. The objective in selecting one company from each level in the automotive supply chain was to analyze the different difficulties experienced in the supply chain. The work was not limited to these four case studies. To include the perspectives of other relevant stakeholders in the process of adoption of IATF Standard, four interviews were conducted with a similar structure (two with ISO/TS 16949 auditors and two with ISO/TS 16949 consultants). The field work started in January 2017 and, as stated, was not finished when the preliminary findings reported in the present chapter were completed in May 2017. A series of two or three semi structured (but very open) in-depth interviews were conducted in each company with managers and technicians with responsibility for the management system. For reasons of confidentiality agreed upon with the companies included in the study, reference is only made to the Tier in which the company operates, as shown in Table 1.

**Table 1** General information of the case studies

| Automotive supply chain level | OEM | Tier 1 | Tier 2 | Tier 3 |
|---|---|---|---|---|
| Number of employees | 12,671 | 575 | 400 | 13 |
| Activity | Assemble and make cars | Production of suspensions of cars | Wire assembly for cars | Paint and surface protection of the body |
| Date of first ISO/TS 16949 certification | Not have (directly VDA 6.1.) | 2002 | 2015 (2010 other factory in the group) | Only ISO 9001:2015 certificate |
| Expect to adopt the new IATF 16949 | – | May 2018 | During 2018 | September 2017 |
| Main motivation to adopt ISO/TS 16949 | – | Customer requirement | Customer requirement | Customer requirement |
| Experience with meta-standards | Very high (ISO 9001, ISO 14001, OSHAS 18001, ISO 50001) | Very high (ISO 9001, ISO 14001, OSHAS 18001, ISO 50001) | High (ISO 9001, ISO 14001, OSHAS 18001) | Low-medium (ISO 9001) |

*Source* Elaborated by the authors

# 5 Results

## 5.1 *Brief Description of Cases*

### 5.1.1 OEM Case

The manager Responsible for Supplier Audits (RSA) thinks that ISO 9001 is excessively basic and general. In his words, "Having adopted IATF 16949, the Tiers have the advantage of selling in different countries and for different customers with just one certification. Before, they had to have one for each. In our case, having ISO 9001 and VDA 6.1. is enough because the last one is harder than IATF and our customers are the users of the cars". Moreover, the manager Responsible for Quality Management Systems (RQMS) says that IATF is a rather general management norm, but he thinks that VDA 6.3. enters fully in the process and is much closer: "We have also audits of VDA 6.3. by the group".

In words of the RSA, "For us it is critical that our customers control their customers. It is not enough to tell us that they do; they must prove it. ISO 9001 does not require such things. It is essential that they demand or communicate our demands with certainty". In his opinion most of the problems originate downstream (Tier 2, 3, N) and that is why you have to have a robust control in all of the supply chain. In addition, their customers (Tier 1) must pass the group Formal Q audit, which consists of VDA 6.3 requirements plus other specific requirements. In his words, it is far away from ISO 9001 and with this they are sure of having a robust process in the supply chain.

### 5.1.2 Tier 1 Case

In words of the manager Responsible for ISO 9001 and ISO/TS 16949 (RII) the main difference between ISO 9001 and ISO/TS is that, "ISO 9001 is too general in relation of services or process management. ISO/TS takes another step in the validation of product design, warranties, and piece returns from the customer (OEM). How do you analyze that? How do you avoid this defect in the future?" But she stated that for customers this is not enough: "All OEMs impose Customer Specific Requirements (CSR) on us, such a VDA audit, and this is harder than the ISO/TS audit. Each OEM wants some requirement in detail, more in depth". To achieve this level, she thinks that it is essential to have ISO/TS as a base, but perhaps ISO/TS should be even more robust.

The RII is sure that they could work without ISO/TS nowadays, but she highlights that "ISO/TS has been very important on the way towards internalize the mode of working in processes at a high level". Sometimes, they have thought that the level was too high, but it helps them to improve. But it is true that sometimes she thinks that they are wasting time because the standard requires indicators of efficiency and efficacy in all process and she thinks that it is not logical because the measures are not applicable in all the processes.

The manager Responsible for Management Systems (RMS) and the RQMS affirm that the level of demand is the same if the factory (one of the group) is ISO/TS or not, because they have internalized the management system. However, there are big differences between their suppliers. For instance, “It is possible to see if it is certified with ISO/TS or not, when they reply in 24 h as a maximum, when they carry out a cause analysis in just one week or when they understand the complexity or necessity of what we demand”. He stresses that, “It is not enough if you send 5 bad products sending 100 new ones. You have to ensure that this is not going to be repeated. It is another way of thinking”.

### 5.1.3 Tier 2 Case

For the RQMS the main difference between ISO 9001 and IATF is the need always to be on top in all the areas because they must be certified each year. ISO 9001 only requires the review of some specific aspects of the management system each year, and a complete review of the management system each three years. This culturally requires you to work with another mentality. The RII1 added that often the goal is to have everything updated all the time and not think about the audit. The RII1 believes that they work more or less at that level so that they could pass an audit on any day of the year. The RII2 affirms that it is daily work but that it does not only depend on them. Sometimes they must pressure the customers and suppliers to have all the documentation updated when they are late with the documents.

The RII1 comments that they work to control as far as possible and they know that they comply with ISO 9001 and ISO/TS. The RQMS adds that any workshop in the group that does not have ISO/TS uses the same documents, QS and processes that they use. They highlight that ISO/TS expects many more concrete things and they work to be the best possible. If a component has failed and this is not permitted, they always seek to find out whether it was a failure in the ISO/TS system. In the words of RQMS, “In terms of flexibility and rigidity, there is not much difference between ISO 9001 and IATF”. Despite this general claim, other specific points addressed by the interviewee contradicted that claim, and he claimed that ISO/TS demands something more than ISO 9001. The RQMS underlined that ISO/TS is the requirement to start working with some specific automotive groups, but then, the OEMs want to formalize their systems. In their case, the ISO/TS does not give them anything in terms of management. It would be enough for their Management System to pass the VDA audits. However, it is necessary to have the certificate to look for new customers since it is a “minimum” requirement to start playing. The RQMS adds that having ISO/TS does not ensure that you pass a VDA audit.

### 5.1.4 Tier 3 Case

This is a particular case because the company was burned down in August 2012, and in January 2014 it almost lost its ISO 9001 certificate. The RMS comments that

ISO 9001 is not enough because customers want them to paint well and to have processes under control and stable. She explains that the old ISO/TS provides for the reduction of variation and wastage. Customers do not want surprises in the process. Both the RQM and the RII say that if the client had not demanded it, they would not have controlled the process that way. To them, ISO/TS is stimulating growth. They confirm that the adoption process is tough but very positive because they are much more competitive. They have beat off competitors from above, and they have changed the shop to work at a very high level. Also, they have been able to enter into other sectors outside automotive manufacture.

The RMS affirms that, "ISO/TS makes us think of the risk in the process. You really have to think about where your risks are, what your contingencies are, how you manage them and if it helps you noticeably improve the process". She added, "More than improving by improving, really seeing where the process can go out of control and really seeing where you can pull a piece. This is not just about conforming to customer demands". The RII adds that the most difficult thing about the standard is that it is not like ISO 9001, even the 2015 version, which is very easy to internalize because it does not have any kind of complicated technical words. In the IATF, you need a lot of training. The RMS has been training for a year because they ask for statistics, very concrete things, that without training and support are very difficult to internalize. The RQM says that the hardest point is that you have to take into account the CSR, which are apart from the IATF, and the customer can demand more of you. They say that, having the IATF, they will suffer less in audits, because they will already be in the routine with a standard and having small variations. On a humble scale, they gradually adapt to the IATF way. They believe that the technology required is a higher level than for ISO 9001. RII gives the example of the in situ data collection. They have to do it on paper and the workers are hindered by having to fill out paperwork, so it will be very beneficial when they can afford to investment in technology.

## 5.2 *Brief Description of Direct Stakeholders*

### 5.2.1 Auditors

Auditor 1 has been an internal auditor of ISO 9001 and ISO/TS 16949 in an OEM company and external auditor for Bureau Veritas and SGS. Auditor 2 has worked for several certification companies as a freelance external auditor of ISO 9001 and ISO/TS 16949 for several years, mainly auditing small companies. Both auditors comment that small companies in the automotive sector that do not have ISO/TS 19949 do not value ISO 9001. For them, ISO 9001 is too simple. An ISO/TS auditor must spend many more hours to carry out an audit because they need more evidence.

The auditors emphasized that ISO 9001 is not accepted in the market as much as ISO/TS, because the ISO/TS chapters, statistical requirements and evidence are much more demanding. Auditor 1 gives a typical example: "The order and cleanliness (…) the ISO/TS requires that you need evidence that the production process is kept clean, not only that the customer does not complain as in ISO 9001. It is necessary to explain how it is kept clean". Auditor 2 emphasizes that often small companies want to implement the ISO/TS thinking that it is filling out more paperwork, and it is not. He explains that it changes the concept of the company and the requirements of the systems are radically higher. "It has to change the mentality of those who are above and it changes the concept of production and treatment of non-conformity… It is necessary to have a more structured system and to have evidence. It is not enough to think that 'it will turn out well', because SPC concepts appear clearly in ISO/TS".

Auditor 2 talks about that being an auditor for ISO 9001 or ISO/TS are very different. He underlines that there is even more difference than between the standards. Similarly, the charges per hour are more than the double and the requirements of the audit are stricter. Auditor 1 states that he has had companies that were willing to accept non-conformities, because they thought that would help them to improve. However, other companies are the opposite. These companies avoided non-conformities looking for certification. In the audits of ISO/TS, "In the end, when the auditor gives the company the final report, the whole Board of Directors was sitting in front of him and they talked about how the company was. This does not happen in the ISO 9001 audits".

### 5.2.2 Consultants

The consultants interviewed have long experience in helping companies to be certified with various management systems. For example, they think that, "In ISO 9001 we can talk about raw material input control. In IATF they will not only control raw materials but also need to know the total traceability. Not only a visual inspection of those materials. And also it is necessary to make more quantitative and qualitative inspections and to have it all registered".

Moreover, one consultant affirms that to certify in IATF, it is impossible for the company to prepare the papers to pass an audit in the last week, as is done sometimes for ISO 9001, because of the number of papers that would have to be filled in. Even at computer level, it would be impracticable. It provides the company with a higher level of control of production, of the costs and at the corporate level. The company is much more controlled and they have much more capacity to detect opportunities for improvement. In a system based on ISO 9001, it is sometimes difficult to find opportunities, and even if they are identified, it is not compulsory to act on them and sometimes companies do not start improvement processes.

Apart from having a greater control, a big value of ISO/TS is that you are able to work for automotive customers and in this sector it is much easier to maintain customers because the ISO/TS is a very important obstacle to market entry for new competitors.

### 5.3 *Summary of Evidence from the Cross-Case Analysis*

For companies that want to adopt the ISO/TS standard, the requirements that they have to internalize all over the company are complex and it takes a long period of development and internal work. The companies need to change their mentality. It is not enough to have no non-conformities, because firms have to ensure that errors cannot recur. Moreover, the customers have to be convinced that they can control all risks, and if something occurs, they have all the required tools to avoid non-conformities.

It is clear that, for companies that achieve a high level of internalization, ISO/TS converts into a daily routine and it is not so complicated to implement. In addition, when the internalization level is high, there is no difference for them between the areas in which they have adopted ISO 9001 or ISO/TS. For these companies, the Customer External Requirements part is the hardest point of ISO/TS 16949 standard and a major difference from ISO 9001. They think that sometimes the demands are too high, to the extent that they are not useful. In other words, they think that the customers want too many things, when less would be enough, especially when the customers are also going to carry out a specific audit.

## 6 Discussion and Conclusions

In the case-studies some relevant flaws and weaknesses of the adoption of ISO/TS 16949 were found. Nevertheless, most of the analysis, based on exploratory qualitative field work, was focused on a comparative analysis of ISO/TS 16949 and ISO 9001. The stakeholders who were interviewed compare ISO/TS with the main global reference standard for quality management, ISO 9001. In general terms the stakeholders stress that ISO/TS 16949 demands much more effort than ISO 9001. It can be said that ISO 9001 is too easy in comparison with ISO/TS and less consultant support is needed, in the light of the preliminary conclusions from the field work. All the managers and other stakeholders agree that ISO/TS 16949 requires much more thought, indicators and control than ISO 9001. They added that it is necessary to change the concept of the management system, internalizing the requirements of the standard in their daily work. After this step, the companies find more opportunities to improve their work using technological and management tools to avoid non-conformities. It is interesting to note how the firms work. For example, the standard does not recommend how to control systems, but the Tiers need evidence of their control, the control of their suppliers and the actions to avoid non-conformities. For this reason, each company must control their processes and their suppliers' processes. It could also be interesting to know if there are differences between these processes and the processes that are outside the scope of this certification.

In the automotive industry ISO 9001 does not seem to add much value for companies. Customers pressure their suppliers to adopt and certify a management system based on the ISO/TS. This is the first step for many companies to have the opportunity to sell their products in the automotive sector. This step is not easy, because of the harder requirements of the standard compared with ISO 9001. The OEMs require that their suppliers have an advanced Quality System based on the specific requirements of the sector. For this reason, suppliers are forced to pass specific customer audits after being certified with ISO/TS. This aspect could be a competitive advantage because after that, it is quite easy to maintain the customers.

As has been shown in the ongoing field work, further research should focus on shedding light on the added value of ISO/TS 16949 and IATF 16949 over ISO 9001. Similarly, the analysis of possible differences between the adoption and certification of ISO/TS 16949 and the new IATF 16949 could be also interesting, considering the remarks made by the auditors in the interviews. Among many other things, it would be interesting to analyze whether the new standard is capable of reducing the gap that currently exists between VDA 6.3. or IQL12 audits and ISO/TS audits, which in the opinion of the interviewees are less demanding than the specific standards of the automotive sector. If the new standard achieves that, it could definitely be a facilitating tool for firms in the supply chain of the automotive sector.

**Acknowledgements** This work has been carried out for the Research Group funded by the Basque Government (Grupos de Investigación del Sistema Universitario Vasco; IT1073-16).

## References

Antonelli C (1994) Localized technological change and the evolution of standards as economic institutions. Info Econ Policy 6(3):195–216

Bevilacqua M, Emanuele Ciarapica F, Giacchetta G, Marchetti B (2011) Implementation of a quality procedure based on Delphi method and the ISO/TS 16949: 2009 in the production of stainless steel tubes for automotive exhaust systems. Int J Qual Reliab Manage 28(8):841–866

Brunsson N, Jacobsson B (2000) The contemporary expansion of standardization. In: Brunsson N, Jacobsson B (eds) A world of standards. Oxford University Press, Oxford, pp 1–17

BSI (2017) Moving from ISO/TS 16949:2009 to IATF 16949:2016, transition guide. http://www.bsigroup.com

Cauchick Miguel PA, Leal AF, Da Silva IB (2011) Implementation of ISO TS 16949 towards business excellence–results from a case study. Int J Bus Excell 4(3):283–297

Eisenhardt K (1989) Building theories from case study research. Acad Manage Rev 14(4):532–550

Heras-Saizarbitoria I, Boiral O (2013) ISO 9001 and ISO 14001: towards a research agenda on management system standards. Int J Manage Rev 15(1):47–65

Hoyle D (2005) Automotive quality systems handbook: ISO/TS 16949: 2002 edition. Butterworth-Heinemann, London

Hys K (2014) Tools and methods used by the Polish leading automotive companies in quality management system. Results of empirical research. J Achiev Mater Manuf Eng 63(1):30–37

Hys K (2015) ISO/TS 16949 analysis of the current trends. Zarządzanie I Finanse, J Man Fin 13 (2):37–45

IATF (2016) Internal documentation. http://www.iatfglobaloversight.org/ISO. ISO Survey
ISO (2016) The ISO survey of ISO 9001 and ISO 14001 certifications. ISO, Geneva
Kartha CP (2004) A comparison of ISO 9000: 2000 quality system standards, QS9000, ISO/TS 16949 and Baldrige criteria. TQM Mag 16(5):331–340
Laosirihongthong T, Dangayach GS (2005) A comparative study of implementation of manufacturing strategies in Thai and Indian automotive manufacturing companies. J Manuf Syst 24(2):131–143
Lupo C (2002) ISO TS 16949 the clear choice for automotive suppliers. Qual Prog 43(3):44–49
Ostadi B, Aghdasi M, Kazemzadeh RB (2010) The impact of ISO/TS 16949 on automotive industries and created organizational capabilities from its implementation. J Ind Eng Manage 3 (3):494–511
Punnakitikashem P, Laosirihongthong T, Adebanjo D, McLean MW (2010) A study of quality management practices in TQM and non-TQM firms: findings from the ASEAN automotive industry. Int J Qual Reliab Manage 27(9):1021–1035
Singh N (2014) Automotive industry response to its global QMS standard ISO/TS-16949. In: Das K (ed) Globalization and standards. Issues and Challenges in Indian Business, Springer, Berlin, pp 121–142
Uzumeri MV (1997) ISO 9000 and other metastandards: principles for management practice? Acad Manage Exec 11(1):21–36
Vaxevanidis NM, Krivokapic Z, Stefanatos S, Dasic P, Petropoulos G (2006) An overview and a comparison of ISO 9000: 2000 quality system standards with related automotive ones (QS9000, ISO/TS16949) and TQM models (MBNQA and EFQM). Organization 4:5
Yin RK (2003) Case study research: Design and methods. Sage Publications, Thousand Oaks

# Implementing the ISO 50001 System: A Critical Review

**Alexander Yuriev and Olivier Boiral**

**Abstract** Published in 2011, the ISO 50001 system on energy management has since become one of the fastest-growing ISO management standards. However, its efficacy, implementation and integration within enterprises have been overlooked in the literature. The aim of this chapter is to analyze the relevance of the ISO 50001 standard to reducing energy consumption, and identify its limitations, advantages, and possible weaknesses. Based on a review of the existing literature on ISO management standards and energy management, it appears that ISO 50001 has inherited advantages and limitations similar to other ISO management systems. This chapter can help researchers and managers to analyze the expected outcomes and success factors of ISO 50001. It also sheds more light on its impact on existing energy-related policies.

**Keywords** Energy management · ISO 50001 · Energy conservation · Energy management system · Environmental management system

## 1 Introduction

According to the International Energy Agency, the overall augmentation of energy demand in the next 25 years might reach as high as 37% (IEA 2014). This rise has motivated many countries to implement incentives, or to mandate market-based certification programs for energy management (Du Plessis 2015). These initiatives are assumed to improve organizational greening and to reduce $CO_2$ emissions (Siciliano et al. 2015).

A. Yuriev (✉) · O. Boiral
Department of Management, Faculty of Business Administration,
Universite Laval, Pavillon Palasis-Prince, Local 1638, 2325,
rue de La Terrasse, Quebec, QC G1V 0A6, Canada
e-mail: alexander.yuriev.1@ulaval.ca

O. Boiral
e-mail: Olivier.Boiral@mng.ulaval.ca



I. Heras-Saizarbitoria (ed.), *ISO 9001, ISO 14001, and New Management Standards*,
Measuring Operations Performance, https://doi.org/10.1007/978-3-319-65675-5_9

One of the main certifiable tools in this area is the international standard for energy management systems (EnMS), ISO 50001. Published in 2011, this standard has since become one of the fastest-growing ISO management standard, with 6778 certificates issued by March, 2014 (ISO 2014). It "enables organizations to establish the systems and processes necessary to improve energy performance, including energy efficiency, use and consumption" (ISO 2011: v). Although a few empirical studies have demonstrated the overall satisfaction of certified organizations with the standard's principles and its economic benefits (Ates and Durakbasa 2012; Gopalakrishnan et al. 2014; Therkelsen et al. 2013), the ISO 50001 system has been under-studied from a critical perspective.

ISO 50001 is frequently implemented alongside the existing environmental (ISO 14001) and quality (ISO 9001) management systems which are based on similar principles (Ates and Durakbasa 2012; Karcher and Jochem 2015; Wulandari et al. 2015). While various aspects of ISO 14001 and ISO 9001 have been found to positively influence operations and management of certified companies (Prajogo et al. 2012; Beattie 1999; Lo et al. 2009; King and Lenox 2009), certain scholars have raised concerns about the duration of these benefits and the very nature of the reasoning behind certification (Rondinelli and Vastag 2000; Christmann and Taylor 2011; Boiral 2007). Many studies point to various limitations and drawbacks related to the Plan-Do-Check-Act (PDCA) and formulaic approach that underlies the ISO management systems, including ISO 50001 (e.g. Castka and Corbett 2015; Yin and Schmeidler 2009; Boiral 2011; Jiang and Bansal 2003; Balzarova and Castka 2008).

These debates about ISO management systems raise questions on the added value of ISO 50001. The objective of this chapter is to analyze the relevance of ISO 50001 to reducing energy consumption, and identify its limitations and advantages. The existing literature on this standard is fairly scarce and consists of only a few articles. In order to back up our discussion with relevant facts, the principles of ISO 50001 were systematically compared with those of ISO 14001 and ISO 9001. Then, the applicability of the concerns related to environmental and quality management systems to ISO 50001 was assessed and analyzed.

The rest of the chapter is organized as follows. First, we review different approaches to energy conservation and explore the evolution of this concept. Second, we identify and explain the fundamental tenets of ISO 50001. Third, we provide a detailed comparison of three management norms. Finally, we proceed to the identification and further exploration of how empirical studies of ISO 14001/9001 can be relevant to ISO 50001. The chapter concludes by suggesting a number of possibilities for future research, together with managerial and policy implications.

# 2 The Principles of ISO 50001

## *2.1 The Development of Energy Management Practices*

Over the last few years, organizations have prioritized one-time cutting and fast transformations in order to decrease their energy consumption. Since the late 70s, managers have reduced the power intake by implementing practically effortless, and low-cost solutions such as turning off unnecessary lights and adjusting air conditioning (i.e. Lambert and Stock 1979; Palm 2009; Introna et al. 2014). The same tendency can be observed with regards to other environmental issues, for example, water conservation (closing tap) or waste management (plastic recycling). Such measures do not require the construction of any system, and can be implemented in any organization, frequently without additional capital investment.

Contrary to this, rather straightforward, approach to energy conservation, some scholars have proposed the concept of "energy management", which "involves monitoring, measuring, recording, analyzing, critically examining, controlling and redirecting energy and material flows through systems so that the least amount of power is expended to achieve worthwhile aims" (O'Callaghan and Probert 1977, p. 128). Although the existing literature does not provide a single definition of energy management (Testa and Vigolo 2015), most of them imply analysis, reporting, and action, all of which together lead to continuous improvement (i.e. Abdelaziz et al. 2011; Capehart et al. 2012), which means that a certain systematic approach is required.

While environmental pressures are increasing, no study has reported a change in the preferences of enterprises to energy conservation. In fact, at the emergence of energy management concept, certain organizations had already attempted to integrate a profound EnMS into their activities. For example, McClelland and Cook (1980) looked at the case of a university in the USA where the administration began the energy-reduction program with quick-fixing measures, gradually advancing it by implementing more sophisticated solutions: energy supervisors, regular check-ups, sensitization of students. Although many organizations still use one-time fixing measures (Introna et al. 2014), more and more organizations adopt systematic approaches, particularly the ISO 50001 system, which has been adopted by almost 12,000 enterprises (ISO 2015a, b, c).

## *2.2 The ISO 50001 Management System*

Analogously to ISO 14001 and ISO 9001, the ISO 50001 energy management standard does not differentiate between organizations with diverse geographical, cultural or social conditions. The criteria for energy assessment form the logic behind the norm and include multiple aspects, the most important of which are: the origin of the current energy consumption; analysis of collected data; setting the energy baseline and performance indicators; establishing objectives for reducing the power intake; documentation control; and continuous improvement.

Figure 1 illustrates the main principles of the ISO 50001 system. These principles are integrated within the PDCA cycle which is also in the heart of ISO 9001 and ISO 14001. In fact, implied steps are almost identical to those suggested by other management standards. The main difference lies in the introduction of so-called "energy baseline" and "energy objectives" in the "Plan" stage of the implementation process. In case of the energy management norm, power reduction actions are chosen based on these two calculations, and therefore a certain definite improvement has to be continuously achieved.

As for the "Do", "Check" and "Act" parts of the system, they are based on principles similar to those of other ISO management systems: top management commitment, appointment of an energy management representative, establishing energy policy, setting an energy baseline, fixing energy objectives and targets, ensuring employee awareness and providing workers with necessary trainings, and, finally, procuring continuous improvement. In fact, official texts of ISO 50001:2011, ISO 9001:2015 and ISO 14001:2015 do not differ significantly (see "Appendix"). The relevance of the above mentioned key practices has been widely covered in the literature on ISO management systems and is summarized in Table 1. Where possible, papers exploring energy-related issues are inserted as reference. Otherwise, relevant studies on quality and environmental management standards related to the practices mentioned are indicated.

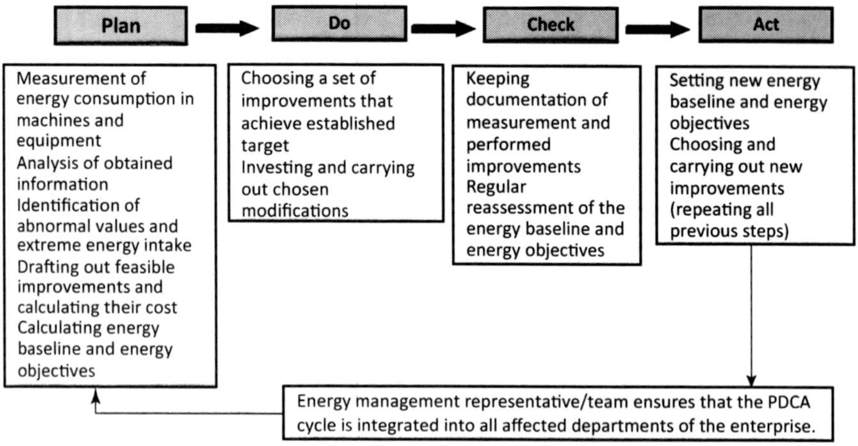


**Fig. 1** PDCA cycle of ISO 50001. *Source* Developed by authors based on the official text of ISO 50001:2011

**Table 1** ISO 50001 principles and evidence from the literature

| ISO 50001:2011 principles | Evidence from the literature | Authors |
|---|---|---|
| 1. Top management commitment (p. 5) | There are three levels of top management commitment to corporate social responsibility: minimum legal compliance; enlightened self-interest; pro-active change | Stahl and Grigsby (1997) |
| | Top management commitment has a significant positive effect on productivity | Rodgers and Hunter (1991), Switzer et al. (1999) |
| | Although the link is weak, organizations conform better to the environmental standards when top management is more committed to them | Chung et al. (2005) |
| | Management commitment indirectly influences environmental behavior at work | Fernández-Muñiz et al. (2012) |
| 2. Energy management representative(s) (p. 6) | Appointment of energy management representatives is an efficient way to reduce energy consumption | McClelland and Cook (1980) |
| | Management representative contributes significantly to the long-term total quality goals of the company | Lo and Sculli (1998) |
| 3. Energy policy (p. 6) | The depth of policy reflects the degree of environmental management system success | Chung et al. (2005) |
| | Environmental policy is the core of environmental management system | Darnall and Kim (2012) |
| 4. Energy baseline (p. 7) | It is essential to estimate the baseline of current energy use in order to set targets for energy reduction | McClelland and Cook (1980) |
| | The success of the restoration ecology can only be assessed by having in mind the exact starting point | Hobbs and Harris (2001) |
| | Greenhouse gas emissions baselines helps to assess the reduction potential | Ådahl et al. (2004) |
| 5. Energy objectives and targets (p. 8) | The notion of monitoring and targeting has first been suggested to conserve energy in the British sheet board, paper and textile industry | Gotel (1989) |
| | The effectiveness of goal-setting for achieving energy efficiency is beyond controversy | Harris (1989) |
| | Various software for energy management integrate the principle of creating objectives | Hooke et al. (2003) |
| 6. Employee awareness and training (p. 8) | Transforming energy-using behaviors is one of the ways to save energy | McClelland and Cook (1980) |

(continued)

**Table 1** (continued)

| ISO 50001:2011 principles | Evidence from the literature | Authors |
|---|---|---|
| | Human errors influence power plant environments, and their correction can lead to significant decrease in energy consumption | Worledge (1992) |
| | The long-term competitive advantage can only be achieved by investing in training of employees | Yiu and Saner (2005) |
| | Training increases the awareness of environmental issues, and helps organizations to achieve its "greening" goals | Sammalisto and Brorson (2008) |
| 7. Documentation control (p. 9) | Efficient documentation control makes the established procedures more accessible and available for employees | Epstein and Roy (1997) |
| | Systematic verification of policies improves the overall quality of documentation and makes environmental management "system dependent" rather than "person dependent" | Morrow and Rondinelli (2002) |
| | The success of the environmental management system depends on many factors, one of which is documentation control | Sambasivan and Fei (2008) |
| 8. Continuous improvement (p. 12) | Continuous improvement is vital to the survival of enterprises | Kaye and Dyason (1995) |
| | Efficient quality management cannot be achieved with "quick fixes" and requires a particular focus on continuous improvement | Kanji and Asher (1996) |
| | Only by integrating the continuous improvement framework can organizations adapt quickly to changes | Kaye and Anderson (1999) |
| | Companies following continuous approach for quality management achieve greater results | Terziovski and Power (2007) |

## *2.3 Implementation and Outcomes of the Standard*

The outcomes of ISO 50001 have been investigated in few recent case studies. According to Gordić et al. (2010), the implementation of the standard resulted in approximately 25% reduction in energy consumption within a Serbian car manufacturer. Similarly, Majernik et al. (2015) verified the effectiveness of ISO 50001 within a car supplier, and confirmed a significant positive change. Imel et al. (2015) has also reported a significant decrease (39%) in energy use within a county in

Florida that integrated multiple principles of the norm into its daily operations. However, this transition might be seen challenging for some companies. Generally speaking, the benefits of any environmental management system—whether it is general, like ISO 14001, or more specific, like ISO 50001—are not automatic and straightforward but rather depend on the implementation process (Yin and Schmeidler 2009; Boiral 2011; Jiang and Bansal 2003). This process has been the object of few studies focused on ISO 50001.

For instance, Antunes et al. (2014) propose an Energy Management Maturity Model to guide enterprises in the implementation of ISO 50001. The paper illustrates the integration of the norm through a five-level process, with specific activities applicable to every level. Authors have also investigated common challenges related to these activities, so that managers could attempt avoiding them. The recent paper of Jovanović and Filipović (2016) suggests a different maturity model, and authors claim to have tested suggested levels within certified and non-certified organizations. Other scholars (Lee et al. 2014) developed a so-called six-sigma approach which, similar to Antunes et al. (2014), is also based on five principal steps: define, measure, analyze, improve and control. The article provides several recommendations and a list of most energy-consuming appliances in the manufacturing industry. It is argued that this six-sigma approach can be successfully used "as part of the ISO 50001 implementation" (Lee et al. 2014: 23). Besides that, there have been some attempts to technologically modernize this process, and to introduce special software that would facilitate the implementation of the norm. Gopalakrishnan et al. (2014) suggested a computer program that allows small and medium-sized enterprises to find non-conformities by answering a list of questions related to ISO 50001. Tarasovskii and Petukhov (2014) made a review of existing automated solutions for metal-processing industry, and put forward some ideas of how the time spent by energy managers on paper-work and on the construction of tables and figures can be significantly decreased.

Introna et al. (2014) proposed an energy management maturity model based on five dimensions: knowledge, methodological approach, information system, organizational structure, and strategy. These dimensions are evaluated on the scale of five levels: initial, occasional, planning, managerial, and optimal. For example, there is little benefit in installing a complex computerized measuring program without raising employees' basic understanding of energy management goals. Introna et al. (2014) argue that it is impossible to "introduce advanced management activities before adopting some elementary ones" (p. 115). Whilst the suggested model can help companies to upgrade their energy practices in compliance with ISO 50001, the advantages and limitations of such an advancement remain unclear. Indeed, why should companies consider implementing energy management standard, if its added value to existing environmental management system established by ISO 14001 or other similar norms seems rather dubious?

## 3 Implementing ISO Management Systems: Lessons from Practice

### *3.1 Selection of Articles*

The critical analysis of the ISO 50001 system was based on two sources: the empirical literature on the ISO management systems and a few specific studies on ISO 50001.

Although the literature on ISO 50001 is relatively scarce, the advantages and limitations of ISO management systems have been the object of many studies (e.g. King and Lenox 2009; Poksinska et al. 2003; Boiral and Roy 2007; Balzarova and Castka 2008). In order to choose the most relevant empirical studies on ISO management systems that could shed more light on the ISO 50001 system, the monograph of Castka and Corbett (2015) was used. This work of more than 200 pages provides the most extensive review of the literature on such management systems as ISO 9001, ISO 14001 as well as on other voluntary standards. Authors initially selected 2836 articles which contained a wide range of keywords [e.g. "ISO 9000 (including ISO 9001 and other variations), ISO 14000, ISO 26000, management standard, process standard" (Castka and Corbett 2015: 177)], and subsequently chose several hundred papers for the analysis by applying specific inclusion and exclusion criteria. As the creators of the monograph were not constrained by the number of pages, this overview provides precise and complete information on every selected study.

However, the work of Castka and Corbett (2015) included only those articles that were published before 2014, which means that only a couple of manuscripts on ISO 50001 were reviewed by the authors. For this reason, we launched multiple searches in ABI/Inform, Business Source, IBSS and Web of Science databases, and found a dozen pertinent studies on ISO 50001. In the selection of these articles, we firstly chose papers that mentioned ISO 50001 or Energy Management at least once, and then reviewed their abstracts. Only studies that specifically explored the decision for adopting the standard, the EnMS design, implementation process, follow-up procedures, issues related to auditing or certification outcomes were ultimately retained.

Having reviewed these papers, a table containing arguments about ISO 9001 and ISO 14001 that are directly applicable to ISO 50001 was created (Table 2). The analysis provided by this table is based on the fact that, as it has been underlined in the scholarly literature (e.g. Heras-Saizarbitoria and Boiral; Boiral et al. 2017), the core idea, implementation and audit processes for above-mentioned ISO standards are almost identical.

Therefore, due to a high degree of resemblance between the official texts of ISO 50001 and other management standards (for additional information on this comparison, please refer to "Appendix"), the conclusions of Castka and Corbett (2015) are relevant to shed more light on the possible outcomes of ISO 50001. The Table regrouped the advantages and limitations of ISO 50001 into the four steps of

**Table 2** Applicability of the literature on ISO 9001/ISO 14001 to ISO 50001

| | Main findings of the literature | Applicability to ISO 50001 and managerial implications |
|---|---|---|
| Certification decision | Internal motivations lead to better internalization of management standards. As a consequence, environmental performance of enterprises is ameliorated. (Prajogo et al. 2012; González-Benito and González-Benito 2005; Boiral and Roy 2007; Terziovski and Power 2007; Naveh and Marcus 2005) | The resemblance of standards makes it possible to suggest that internalization of ISO 50001 might be greater when motivation for the certification originates inside the organization. It might also result in better energy conservation. Therefore, environmental managers of companies are encouraged to prepare the organization to the installation of energy management system by explaining its importance and the multitude of advantages |
| | Higher motivation for management standards leads to better financial results (Martínez-Costa et al. 2008; Montalván and Chang 2006; Jacobs et al. 2010; Lo et al. 2012; Nicolau and Sellers 2002) | Implementing ISO 50001 leads to increased income. Companies that participated in the survey of Therkelsen and McKane (2013) saved on average 503,000$ in less than 2 years. In our opinion, enterprises with significant energy consumption might find this figure motivating, which will result in a more in-depth integration of the standard |
| | Adoption of management standards leads to organizational learning. As a consequence, implementation of the second or third norm, based on the same documentation (e.g. ISO 9001 and ISO 14001) occurs more rapidly (Karapetrovic and Casadesús 2009). Also, the adoption of one management standard in many cases predicts the willingness of companies to integrate other ones (King and Lenox 2009) | EnMS is usually built on already existing management structures (Karcher and Jochem 2015; Wulandari et al. 2015). Hence, the adoption of ISO 14001 might encourage and facilitate the implementation of ISO 50001. In fact, we argue that companies that have already been certified with ISO 14001 might find the energy management standard redundant. Considering the similarity between these norms, it seems reasonable for enterprises to integrate energy-related issues inside previously created ISO 14001 system |
| | In organizations driven by external pressures, the certification process tends to be shaped by the degree-purchasing syndrome (Boiral 2012) | External pressures for the adoption of ISO 50001 tend to encourage a symbolic adoption of the standard and the search for certification as an end in itself. It is recommended that companies adjust their certification-acquiring process according to its proper motivations. |

(continued)

**Table 2** (continued)

| | Main findings of the literature | Applicability to ISO 50001 and managerial implications |
|---|---|---|
| | | The degree of implementation, the involvement and training of employees, the purchase of energy saving equipment, precision in measuring energy intake—the degree of integrating all these aspects should be carefully considered in relation to companies' willingness, and to external pressures |
| System design | Management standards should be customized to the needs of a particular company in order to be integrated efficiently (Peter 1995; Naveh and Marcus 2005). Motivations for acquiring management certifications should determine the procedures and documentation, not vice versa (Bénézech et al. 2001) | According to some scholars (e.g. Westphal et al. 1997; Attewell 1992), the process of standard's optimization is best managed by trained employees of the organization, and not by external consultants. Externalization of these processes might make the implementation more burdensome, and less effective. It might be pertinent for EnMS managers to consider forming a system-customization team, in order to better integrate it within the company |
| | In many instances companies decide themselves what to make of standards (Boiral 2011) | Managers might attempt to integrate the principles of ISO 50001 into the existing management system, with the purpose to reduce administrative procedures and paperwork. However, such adjustments should be carried out with caution, and should not have a negative effect on the efficiency of the system |
| Implementation | ISO management systems can be considered as a tool to transfer individual and collective knowledge (Bénézech et al. 2001; Boiral 2011) | ISO 50001 can facilitate knowledge transfer on energy management: documents are created to organize and manage the existing and evolving information more efficiently. However, managers are encouraged to minimize the amount of unnecessary paperwork by merging similar or identical procedures in case of several management systems (e.g. quality and environmental ones) |
| | The proper adoption of management standards depend on employees | As it is the case with all management systems, the involvement of |

(continued)

**Table 2** (continued)

| | Main findings of the literature | Applicability to ISO 50001 and managerial implications |
|---|---|---|
| | (Darnall and Kim 2012; Balzarova et al. 2006) and managers (Van der Wiele et al. 2009; Boiral 2003, 2011; Poksinska 2007). Their greater involvement lead to more profound implementation of norms' principles (Fuentes et al. 2003; Poksinska 2007; Boiral 2003) | employees is the key to a successful implementation of ISO 50001. Managers should contemplate various types of employees' involvement: launching specific initiatives, encouraging workers' endeavors, introducing new policies addressing in-role energy reducing activities (for example for operators working with high energy-consuming equipment), favoring voluntary behaviors (turning off lights, adjusting air-conditioning, reducing phantom energy, etc.) |
| | ISO management systems are certified on average within 12 months (Singh et al. 2006) | A study by Karcher and Jochem (2015) indicates that 59% of companies aim to be certified with ISO 50001 within 6 months. This is achieved with a maximum delay of half a month. However, companies should bear in mind that rapid adoption of the standard might diminish its positive outcomes |
| Audit process | Companies certified with ISO 14001 by reputed registrars comply better to environmental policy statements (Fryxell et al. 2004). Some enterprises choose better-known auditors as they tend to give more attention to details, which might lead to additional improvements in the future (Castka et al. 2015). Effectiveness of auditors varies (Gyani 2008; Lafuente et al. 2010), and even the interpretation of such terms as continuous improvement is subject to debate (Burdick 2001; Ammenberg et al. 2001). While some registrars primarily conduct audits "for compliance", others perform audits "for continuous improvement" (Poksinska et al. 2006; Power and Terziovski 2007) | The auditing process of ISO 14001 and ISO 50001 is similar. Companies willing to adopt energy management standard should consider the reputation of the audit-performing registrar. Reputed agencies might pay more attention to specific details that could eventually be fixed and bring additional benefit to the whole management system. Managers should also understand in advance whether they seek for certification with the sole purpose of "having it" or with the desire to improve their processes |
| | In some instances, auditors might be influenced by the remuneration system, and their judgments might not be accurate (Dogui et al. 2014). | While it is difficult to minimize the effect of the remuneration system on the external verification, companies undergoing ISO 50001 certification |

(continued)

**Table 2** (continued)

| | Main findings of the literature | Applicability to ISO 50001 and managerial implications |
|---|---|---|
| | This conflict of interest leads to procedural, rather than substantial, audits (Heras-Saizarbitoria et al. 2013) | might provide comfortable conditions for auditors in order to ensure unbiased certification. These include, and are not limited to: rapid and easy access to documents and employees; reduction of audit-related pressures by avoiding conversations on such topics as time, payment, and involved resources; presentation of the situation "as is", without artificial enhancement |
| | Sometimes, lines between consulting and auditing get blurry (Terziovski and Power 2007) | Managers responsible for the EnMS should have a clear understanding of the difference between consultants and auditors. While the first ones assist companies in integrating the norm, the latter ones are responsible for verifying it. If necessary, incompliances might be eliminated in the aftermath of the audit, with or without the help of consultants |
| Follow-up | Firms certified with ISO 14000 are more likely to assess their suppliers' environmental performance and to set certain criteria (Arimura et al. 2011) | According to ISO 50001, companies should evaluate energy consumption of purchased equipment and machines, and provide suppliers with criteria for their selection. The standard recommends these steps only for purchases that can potentially significantly influence energy consumption. In our opinion, managers should not forget about multiple equipment that require low energy intake, but once their impact is accumulated, the consumption might be high: bulbs, air conditioners, computers, printers, etc. |
| | The fact of having a certification does not guarantee continuous improvement of processes and systems (Granerud and Rocha 2011; Boiral 2003; Heras-Saizarbitoria et al. 2013) | Managers of companies willing to obtain ISO 50001, should not rely solely on the audit decision in evaluating the success of the energy management. It is important to ensure that the system functions at all times, and not only during audits (Boiral 2011) |

(continued)

**Table 2** (continued)

| | Main findings of the literature | Applicability to ISO 50001 and managerial implications |
|---|---|---|
| | The process of improvement is not over when the firm becomes certified. The presence of Environment Management System team is necessary to ensure continuous improvement (Balzarova and Castka 2008) | ISO 50001 emphasizes the importance of an energy management representative/team. Riche (2013) argues that energy consumption is an inter-department challenge. Karcher and Jochem (2015) provide empirical evidence that energy management teams include employees from various departments (mostly, production and management departments). Depending on size and structure of the enterprise, managers should form such teams accordingly, including as many employees from relevant departments as possible |
| Outcomes | Few companies experience no effect of management standards on productivity, cost savings, company profitability, and waste reduction (Santos et al. 2011; Lo et al. 2009; Levine and Toffel 2010; Poksinska 2007; King and Lenox 2009; Russo 2009; Darnall and Kim 2012) | Recent studies (Karcher and Jochem 2015; Therkelsen and McKane 2013; Wulandari et al. 2015) have shown that ISO 50001 has a positive financial effect. According to Therkelsen and McKane (2013), the average payback period of ISO 50001 installation is 1.7 years. Without any doubt, this can only be achieved if energy intake is decreased, and hence the negative impact on the environment is minimized. Companies should bear in mind that implementation of energy management system might not only be beneficial from the economic point of view, but might also help to optimize processes, and to construct a responsible environmental image |
| | Companies certified with management standards reduce their emissions more than enterprises that follow the principles of such norms, but are not certified (Alberti et al. 2000; Potoski and Prakash 2005) | Ates and Durakbasa (2012) argue that the more principles of ISO 50001 are taken into account, the better results are achieved. However, the pressure of certification audit might motivate managers to integrate the system more profoundly, and might lead to additional reduction of energy consumption |

successful implementation of ISO Management Standards evidenced by Boiral (2011): certification decision, system design, implementation, follow-up. Two additional categories were added to this classification: audit process and outcomes. While audit process is linked with follow-up activities, the literature on the subject clearly distinguishes internal organizational issues related to continuous amelioration of the system (e.g., Arimura et al. 2011; Boiral 2003; Granerud and Rocha 2011), and problems linked with auditors' capabilities and expertise (e.g., Dogui et al. 2014; Heras-Saizarbitoria et al. 2013; Terziovski and Power 2007). The "outcomes" category was introduced with the purpose to compare results of adopting environment management systems, and to potentially show the limited added value of ISO 50001. Where empirical evidence exists, our suggestions about the possibility of identical advantages and limits of the energy management standard were supported by recent investigations on the norm. Several illustrative examples of the similarity in the arguments about every step of the successful implementation process are further presented and explained.

## *3.2 Certification Decision*

The majority of researches on certification decisions are tightly linked to the nature of motivation that drives companies to proceed with the integration of management systems. When the decision to get certified stems from inside the company (as compared to external pressures) financial indicators tend to improve (Martínez-Costa et al. 2008; Montalván and Chang 2006; Jacobs et al. 2010). The same remark applies to environmental performance (Prajogo et al. 2012; González-Benito and González-Benito 2005; Boiral and Roy 2007). Drawing on these observations, it seems reasonable to assume that internal motivation for EnMS leads to better results than when such a decision is driven by external pressures, including from suppliers or the market. Boiral (2012) argues that when certification is motivated by external pressures, the standard acquisition and the preparation for the audit tend to be shaped by the degree-purchasing syndrome. That means that companies would adopt these norms ceremonially, and there is no guarantee that they will actually follow integrated principles after the certification is granted.

However, it is rare to find an organization that would adopt ISO 50001 before the implementation of more established ISO management standards, particularly ISO 9001 and ISO 14001 (Karcher and Jochem 2015; Wulandari et al. 2015). In fact, most companies use quality and environmental management systems as a base for the introduction of EnMS due to the necessity to integrate similar documentations and procedures. Moreover, Karapetrovic and Casadesús (2009) argue that the implementation of one ISO standard leads to a certain organizational learning, which facilitates the adoption of consecutive similar norms. As the average integration time for EnMS standard is 6 months (Karcher and Jochem 2015), which is significantly less than for other management standards (Singh et al. 2006), organizations seem to be already aware of various principles of ISO 50001.

This observation provides evidence for the similarity between standards, and also raises the question of whether companies should adopt ISO 50001, which we discuss further in the article.

## 3.3 System Design

The process of integrating specific system inside a company has been the object of several studies (Peter 1995; Naveh and Marcus 2005; Bénézech et al. 2001). These studies underline the importance of management systems' customization in order to address particular needs of firms. It is argued that the superficial implementation of ISO management systems tends to reduce their effectiveness (Peter 1995; Naveh and Marcus 2005). One can assume that the same remark applies to the adoption of ISO 50001. Previous investigations (Westphal et al. 1997; Attewell 1992) have shown that systems are better customized by internal employees rather than external consultants. Hence, the creation of a team that would be responsible for the installation of EnMS within the existing environment might help to adjust elements of the system to the specific needs of the company.

## 3.4 Implementation

The involvement of employees appears as the key to successful integration of management standards (Darnall and Kim 2012; Balzarova et al. 2006). Without workers' participation, most systems become ceremonial and ineffective (Yin and Schmeidler 2009; Boiral 2007; Heras-Saizarbitoria and Boiral 2013). Similar to ISO 50001, new versions of quality and environmental management system standards do not suggest clear recommendations for increasing personnel's participation in pro-environmental activities. However, besides regular employees, top managers and middle-managers are also supposed to actively take part in this process (Van der Wiele et al. 2009; Boiral 2003, 2011; Poksinska 2007). Analogously to ISO 9001 and ISO 14001, the ISO 50001 system highlights the importance of top management involvement before, during, and after the process of certification. Hence, managers responsible for the adoption of energy management system should bear in mind the importance of various activities directed at increasing employees' involvement: incentive programs, new policies for operating energy-intensive equipment, promoting voluntary endeavors, etc. Table 2 provides more information on different behaviors that could be beneficial for ISO 50001.

Another important observation in relation to the implementation of management standards is the time it takes from the certification decision, to the reception of the official proof of a successful audit. Singh et al. (2006) argue that certificates for ISO 9001 and ISO 14001 are usually received within 12 months. Interestingly, Karcher and Jochem (2015) found this figure to be almost twice as little for ISO 50001.

Around 60% of companies that took part in their survey got certified within a maximum of 7 months. This decreased time confirms the existence of the organizational learning, which can also raise the question of the added value of the EnMS standard.

## 3.5 Audit Process

It should be noted that the auditing processes for ISO 14001/9001 and ISO 50001 are essentially the same. Auditors of different standards usually represent the same company, and sometimes are trained to verify compliance to multiple norms. However, scholars have questioned for years the interpretation of certain terms by auditors. For example, Burdick (2001) and Ammenberg et al. (2001) raised concerns about how the notion of "continuous improvement" is understood by external inspectors. That is why some articles question the very nature of audits: "for continuous improvement" or "for compliance" (Poksinska et al. 2006; Power and Terziovski 2007).

Another issue lays in the way auditing firms are chosen and remunerated by organizations. Audit companies usually bill their clients on hourly basis (Dickins et al. 2008; Leventis et al. 2005), and, hence, companies expect auditors to perform their duties in the shortest time frames. This might lead to skewed judgments (Dogui et al. 2014), and to procedural audits without substantial logic (Heras-Saizarbitoria et al. 2013). Not only are such issues certainly present in case of ISO 50001, they might also be even worse because of the peculiarity of ISO 50001. Indeed, the integration of EnMS implies multiple measures that auditors might not be aware of, or might not be capable to estimate accurately in a short period of time and under pressure from the organization undergoing certification. In fact, a few published articles discuss the difficulty of proper energy intake evaluation in industrial processes (Giacone and Mancò 2012), the lack of energy data for some organization types (Dzene et al. 2015), and even the necessity to take into account such highly volatile variables as weather conditions and production levels (Lammers et al. 2011).

## 3.6 Follow-Up

ISO management standards are based on the idea of continuous improvement, which does not happen "automatically" after certification (Granerud and Rocha 2011; Boiral 2003; Heras-Saizarbitoria et al. 2013). Yet, some companies might consider that once certified, the system should work without additional efforts. The continuous adjustments and modifications are required at all times, and not just at the moments of audits in order for such systems to work (Boiral 2011). To ensure such improvements, Balzarova and Castka (2008) emphasize the importance of an environmental management team. ISO 50001 also states that energy management team is required for further enhancement of the installed system. Moreover, some scholars have pointed to the necessity of insuring its inter-departmental nature (Riche 2013; Karcher and Jochem 2015). When employees from different departments are united for solving energy-related issues, they produce more efficient results (Riche 2013). Yet, Karcher and Jochem (2015) provide support that most companies compose energy management teams of workers from production and management departments only. We argue that managers of EnMS project should create such teams depending on such factors as size and structure of the enterprise. Perhaps, representatives of those departments that have the most influence on energy consumption should be invited, in order to provide ground for an efficient dialogue and to avoid cumbersome meeting for employees from irrelevant departments.

## 3.7 *Outcomes*

Most studies on ISO management systems' outcomes mention the positive effects of their implementation, including in terms of productivity, cost savings, company profitability, and waste reduction (Santos et al. 2011; Lo et al. 2009; Levine and Toffel 2010; Poksinska 2007; King and Lenox 2009; Russo 2009; Darnall and Kim 2012). The same remark seems to apply to ISO 50001. Recent studies (Karcher and Jochem 2015; Therkelsen and McKane 2013; Wulandari et al. 2015) provide evidence of reduced operating costs and energy intake. It might also be suggested that EnMS has a positive effect on productivity and company's global environmental image, although this point has not been supported by empirical studies. Another series of studies compares outcomes of certified companies with those of firms that attempt to follow principles of the standards, but prefer not to get certification (Alberti et al. 2000; Potoski and Prakash 2005). These articles argue that companies that underwent external audits generally reduce their emissions and environmental footprint more significantly than their non-certified counterparts. The same effect can be expected with ISO 50001, although further research is needed to confirm or deny this hypothesis.

## 3.8 *The Relevance of Adopting ISO 50001*

The similarity between the most widespread environmental management standard, ISO 14001, and ISO 50001 (see "Appendix" for the comparison), as well as multiple common advantages and disadvantages described in the sub-sections above tend to question the relevance of implementing this EnMS. Indeed, why cannot enterprises simply adjust their environmental system to take into account energy consumption, without having the pressure of being audited, and without introducing a whole set of new documentation? Furthermore, as underlined by Zobel (2008), energy, as an environmental aspect, is already included in the scope of the EMSs.

This question seems all the more important that ISO 50001 does not even emphasize the use of renewable energy (Laskurain et al. 2015), which would to some degree justify the necessity of energy standard. Recent research on the renewable energy sources with ISO 50001 certified companies (Laskurain et al. 2015) demonstrated that most benefits can be observed only in those organizations that go beyond the norm's requirements. Considering the fact that the use of alternative energy has been widely supported by the academia (Kaygusuz 2007; Kaygusuz et al. 2007; Sovacool 2009), this observation is an important omission of the norm.

Furthermore, Johnson et al. (2013) conducted a comparison analysis between the Ship Energy Efficiency Management Plan—the standard usually used in the shipment industry—and ISO 50001. After the analysis of two standards, authors found that the EnMS standard appears to be more exhaustive: it includes all the requirements suggested in the maritime document, and recommends additional amelioration. Yet, according to our study, enterprises might achieve the same results by integrating energy aspect in their ISO 14001 system. Managers would simply need to regularly measure the use of energy in the company, in order to ensure constant decrease. As explained previously, ISO 50001 prescribes enterprises to introduce so-called energy review and energy baseline in order to have verifiable statistics on the energy consumption. While the calculation of these data might lead to more quantifiable results and control, it should result in increased amount of documentation. Although interviewees of the recent study conducted by Laskurain et al. (forthcoming) seem to state that "the paperwork load is greater in the case of ISO 14001 than ISO 50001" (p. 13), the number of documents related to various management standards will certainly increase.

It is also worth saying that no standard can cover every aspect of all industries. Some enterprises will always try to either develop their own, more strict, requirements for energy management, or to establish the system of exchanging best practices and know-hows with the companies of their sector (Ates and Durakbasa 2012). However, after the conducted analysis we were puzzled by the lack of added value of the standard, and perhaps further researches could shed more light on this important issue.

## 4 Conclusions and Implications

The aim of this chapter was to analyze the proposals and added value of ISO 50001 and the existing literature on the subject of energy management. The principles of the EnMS standard were compared with such wide-spread international norms as ISO 14001 and ISO 9001, and it was found that they have numerous points in common: the composition of the energy policy, objectives, training and awareness, control of records and internal audit.

The monograph of Castka and Corbett (2015) that provides an extensive review of articles on various management standards was used for the critical analysis. The applicability of this review to ISO 50001 was assessed, and managerial implications were suggested. Several recent empirical papers found during the review of the literature on the EnMS standard were used in this process. More specifically, they made it possible to compare some findings on ISO 50001 with those on ISO 9001/14001 and revealed the absence of significant differences. Many of the findings on quality and environmental management standards might be directly applicable to the energy management norm. Despite our attempts to find advantages and limitations related exclusively to ISO 50001, only a couple of studies found standard-specific limitations. One study compared this norm with regulations in the energy domain, and the other discussed the omission of renewable energies in the standard. These research studies raise more doubts about the usefulness of ISO 50001, as it seems to bring little value to existing environmental management standards, particularly the ISO 14001 system.

## *4.1 Theoretical Contributions*

This chapter provides an up-to-date review of research on the ISO 50001 energy management standard. The majority of these studies considered ISO 50001 to be an example of good practice (Jovanović and Filipović 2016; Antunes et al. 2014; Lee et al. 2014) but did not critically assess the value of the norm in comparison to existing international norms. Our analysis highlights the limited value of introducing ISO 50001 into companies already certified with ISO 14001. It has been hypothesized that common pitfalls of introducing widespread quality or environmental management standards will also be found in the case of ISO 50001. This chapter finds evidence that confirms this suggestion by comparing the results from the literature on ISO 9001, ISO 14001 and ISO 50001. Also, some studies point to the absence or oversimplification of energy-specific features in the standard: renewable aspects (Laskurain et al. 2015), availability of energy data (Dzene et al. 2015), and the need for technical expertise (Karcher and Jochem 2015). All these doubts reinforce our conclusion on the dubious benefits of adopting this standard.

## *4.2 Practical Implications*

Managers interested in energy management should reassess the actual need of their companies with regards to ISO 50001, and attempt to integrate energy aspects in more recognized ISO management systems, particularly ISO 14001. If a company does not have a management system, we would advise it to consider introducing

ISO 14001, as it will potentially take into consideration a larger variety of environmental issues, rather than a specific energy-related standard. However, based on the literature, it is possible to provide a set of recommendations for companies that undergo EnMS certification, and these are set out in Table 2.

This chapter has also practical implications for public authorities. Given the uncertain added value of ISO 50001, the introduction of incentive-based schemes to encourage companies to green their energy-related operations, as has been done in Germany (ISO 2014), seems questionable. While these financial support programs certainly increase the number of certified companies, we doubt they have a significant positive impact on energy performance, although some studies (e.g. Stenqviest and Nilsson 2012) suggest they do. More attention should be paid to directing companies towards substantial rather than ceremonial implementation of existing environmental norms.

Also, considering two standard-specific limits of ISO 50001 (measuring energy consumption and omitted emphasis on the use of renewable energy), our research urges policy-makers to make additional efforts to develop a trustworthy methodology for measuring energy use independent of such volatile conditions as weather or production levels, and to ensure that such important aspects as renewable energy occupy central place in the policy. Ensuring these two points would bring some added value to future versions of ISO 50001, and might have a much bigger impact on the improvement of energy performance of certified enterprises.

## 4.3 Limitations and Future Research

At the time of writing the present chapter, relatively few papers exploring ISO 50001 were found. The comparatively recent date of the norm's publication and the lack of literature on the question made it challenging to critically assess various aspects of the standard's implementation and operation. It would be interesting to repeat the analysis at a later date, including new articles, that will, without doubt, appear in the next few years. It might reveal other dubious points of the norm for companies that contemplate getting certified with ISO 50001.

This chapter suggests the irrelevance of ISO 50001 for companies that are already certified with ISO 14001, but this proposition is made solely on a theoretical basis. It would be extremely important for future research to test the added value of the EnMS standard for other environmental systems. Many angles can be taken into account in such research: an employee perspective, the actual ecological footprint of enterprises, managers' involvement in two environmentally-related standards, and the priorities of the company when similar management systems are being installed. Combining qualitative and quantitative methodologies might help to shed more light on the usefulness of ISO 50001.

Another crucial aspect that needs to be addressed is the preparedness of auditors to verify companies' compliance with the energy management standard. As explained in the chapter, the standard requires some technical knowledge, skills, tools and equipment to carry out a proper certification audit. The lack of these might lead to the increase in certified companies that do not actually comply with the standard's principles. More extensive research is needed to answer this important question with certainty.

## Appendix

Explanations on the table:
This table includes the comparison of ISO 50001:2011 with two international standards: ISO 9001:2015 and ISO 14001:2015. That is why the column "ISO 50001:2011" is the center of this table: it can be viewed from that column to the left (with the purpose to view the comparison of the energy management standard with the quality one), as well as to the right (to compare the energy management standard with environmental one).

The sign signifies important points with certain differences in these standards.

| Comparison | ISO 9001:2015 | ISO 50001:2011 | ISO 14001:2015 | Comparison |
|---|---|---|---|---|
| ISO 9001:2015 is more explicit about the specific requirements for the management system than ISO 50001:2011 in this section | 4.3 Determining the scope of the quality management system<br>4.4 Quality management system and its processes | 4.1 General requirements | 4.3 Determining the scope of the environmental management system<br>4.4 Environmental management system | ISO 14001:2015 includes all general requirements of ISO 50001:2011 and has some additional ones: "organizational units, functions and physical boundaries" (p. 6); "authority and ability to exercise control" (p. 6). This scope should "be available to interested parties" (p. 7) |
| ISO 50001:2011 emphasizes the necessity of "appointing a management representative and approving the formation of an energy | 5.1 Leadership and commitment | 4.2 Management responsibility<br>4.2.1 Top management | 5.1 Leadership and commitment | ISO 50001:2011 emphasizes the necessity of "appointing a management representative and approving the formation of an energy |

(continued)

(continued)

| Comparison | ISO 9001:2015 | ISO 50001:2011 | ISO 14001:2015 | Comparison |
|---|---|---|---|---|
| management team" (p. 5). This statement is absent in ISO 9001:2015 | | | | management team" (p. 5). This statement is absent in ISO 14001:2015 |
| | Not applicable | 4.2.2 Management representative | A.5.3 Organizational roles, responsibilities and authorities | ISO 50001:2011 has more duties assigned to management representative. ISO 14001:2015 only mentions the possibility of the top management to appoint a representative, or a group of representatives to be responsible for environmental policy |
| No significant difference (quality policy is a more specific version of environmental policy) | 5.2 Policy | 4.3 Energy policy | 5.2 Environmental policy | No significant difference (energy policy is a more specific version of environmental policy) |
| ISO 9001:2015 contains a set of specific questions designed to assist organizations in planning. ISO 50001:2011 illustrates the process of planning in a chart (p. 16) | 6.2 Quality objectives and planning to achieve them | 4.4 Energy planning<br>4.4.1 General | 6.2.2 Planning actions to achieve environmental control | ISO 14001:2015 contains a set of specific questions designed to assist organizations in planning. ISO 50001:2011 illustrates the process of planning in a chart (p. 16) |
| | Not applicable | 4.4.2 Legal requirements and other requirements | 6.1.3 Compliance obligations | No significant difference |

(continued)

(continued)

| Comparison | ISO 9001:2015 | ISO 50001:2011 | ISO 14001:2015 | Comparison |
|---|---|---|---|---|
| A peculiarity of ISO 50001:2011 | | 4.4.3 Energy review | | A peculiarity of ISO 50001:2011 |
| A peculiarity of ISO 50001:2011 | | 4.4.4 Energy baseline | | A peculiarity of ISO 50001:2011 |
| No significant difference | 6.2 Quality objectives and planning to achieve them | 4.4.5 Energy performance indicators | 6.2.2 Planning actions to achieve environmental control | According to ISO 50001:2011, energy performance indicators have to be measurable in order to compare the results with the energy baseline. ISO 14001:2015 only requires measurable targets "if practicable" (p. 10) |
| No significant difference | 6.2 Quality objectives and planning to achieve them | 4.4.6 Energy objectives, energy targets and energy management action plans | 6.2.1 Environmental objectives | No significant difference |
| | Not applicable | 4.5 Implementation and operation 4.5.1 General | Not applicable | |
| No significant difference | 7.2 Competence 7.3 Awareness | 4.5.2 Competence, training and awareness | 7.2 Competence 7.3 Awareness | No significant difference |
| ISO 9001:2015 includes more specifications on the communication of information regarding the management system or its results | 7.4 Communication | 4.5.3 Communication | 7.4 Communication 7.4.1 General 7.4.2 Internal communication 7.5.3 External communication | ISO 14001:2015 includes more specifications on the communication of information regarding the management system or its results. It also emphasizes the importance of external |

(continued)

(continued)

| Comparison | ISO 9001:2015 | ISO 50001:2011 | ISO 14001:2015 | Comparison |
|---|---|---|---|---|
| | | | | communication (if necessary) |
| ISO 50001:2011 has more documentation requirements, as data on energy baseline, objectives and review has to be recorded | 7.5 Documented information 7.5.1 General | 4.5.4.1 Documentation requirements | 7.5 Documented information 7.5.1 General | ISO 50001:2011 has more documentation requirements, which can be explained by the necessity to record all information about energy baseline, objectives and review |
| ISO 9001:2015 introduces the "control of changes" requirement for documents (p. 9). Besides that, no significant difference | 7.5.2 Creating and updating 7.5.3 Control of documented information | 4.5.4.2 Control of documents | 7.5.2 Creating and updating 7.5.3 Control of documented information | ISO 14001:2015 introduces the "control of changes" requirement for documents (p. 13). Besides that, no significant difference |
| ISO 9001:2015 sets the requirement to keep documentation when planned operations are carried out in order to ensure the conformity of services to requirements (p. 10) | 8.1 Operational planning and control | 4.5.5 Operational control | 8.1 Operational planning and control | ISO 50001:2011 sets these requirements only for those activities with "significant energy use" (p. 10). ISO 14001:2015 provides examples of controls, and emphasizes that they "can be used individually or in combination" (p. 13) |
| | Not applicable | 4.5.6 Design | Annexes | ISO 50001:2011 requires to consider energy use in the |

(continued)

(continued)

| Comparison | ISO 9001:2015 | ISO 50001:2011 | ISO 14001:2015 | Comparison |
|---|---|---|---|---|
| | | | | planning and construction of new buildings or facilities. ISO 14001:2015 does not include explicitly such requirement, although some annexes do make special notes about buildings [A.1 (p. 18), A.4.3 (p. 21), A.6.1.2 (p. 24)] |
| ISO 9001:2015 provides more specifications on the information to communicate to suppliers regarding its requirements | 8.4 Control of externally provided products and services<br>8.4.1 General<br>8.4.2 Type and extent of control<br>8.4.3 Information for external providers | 4.5.7 Procurement of energy services, products, equipment and energy | 8.1 Operational planning and control | No significant difference |
| ISO 9001:2015 leaves it at the discretion of enterprises to decide when, what and how to monitor advancement | 9.1 Monitoring, measurement, analysis and evaluation<br>9.1.1 General | 4.6.1 Monitoring, measurement and analysis | 9.1 Monitoring, measurement, analysis and evaluation<br>9.1.1 General | ISO 14001:2015 prescribes organizations to set criteria when re-evaluation is needed, ISO 50001:2011 simply states "respond to significant deviations in energy performance" (p. 11) |
| | Not applicable | 4.6.2 Evaluation of compliance with legal requirements and other requirements | 9.1.2 Evaluation of compliance | No significant difference |

(continued)

(continued)

| Comparison | ISO 9001:2015 | ISO 50001:2011 | ISO 14001:2015 | Comparison |
|---|---|---|---|---|
| No significant difference | 9.2 Internal audit | 4.6.3 Internal audit of the EnMS | 9.2 Internal audit<br>9.2.1 General<br>9.2.2 Internal audit programme | No significant difference |
| No significant difference | 10.2 Nonconformity and corrective action | 4.6.4 Nonconformities, correction, corrective action and preventive action | 10.2 Nonconformity and corrective action | No significant difference |
| ISO 9001:2015 emphasizes the importance of protecting documents and their confidentiality and legitimacy | 7.5.2 Creating and updating<br>7.5.3 Control of documented information | 4.6.5 Control of records | 7.5.2 Creating and updating<br>7.5.3 Control of documented information | ISO 14001:2015 emphasizes the importance of protecting documents, their confidentiality and legitimacy |
| No significant difference | 9.3 Management review | Management review<br>4.7.1 General | 9.3 Management review | No significant difference |
| No significant difference | 9.3 Management review | 4.7.2 Input to management review | 9.3 Management review | No significant difference |
| No significant difference | 9.3 Management review | 4.7.3 Output from management review | 9.3 Management review | No significant difference |

# References


Abdelaziz E, Saidur R, Mekhilef S (2011) A review on energy saving strategies in industrial sector. Renew Sustain Energy Rev 15:150–168

Ådahl A, Harvey S, Berntsson T (2004) Process industry energy retrofits: the importance of emission baselines for greenhouse gas reductions. Energy Policy 32(12):1375–1388

Alberti M, Caini L, Calabrese A, Rossi D (2000) Evaluation of the costs and benefits of an environmental management system. Int J Prod Res 38:4455–4466

Ammenberg J, Wik G, Hjelm O (2001) Auditing external environmental auditors—investigating how ISO 14001 is interpreted and applied in reality. Eco Manage Audit 8(4):183–192

Antunes P, Carreira P, Mira da Silva M (2014) Towards an energy management maturity model. Energy Policy 73:803–814

Arimura TH, Darnall N, Katayama H (2011) Is ISO 14001 a gateway to more advanced voluntary action? The case of green supply chain management. J Environ Econ Manage 61:170–182

Ates SA, Durakbasa NM (2012) Evaluation of corporate energy management practices of energy intensive industries in Turkey. Energy 45:81–91
Attewell P (1992) Technology diffusion and organizational learning: the case of business computing. Org Sci 3(1):1–19
Balzarova MA, Castka P, Bamber CJ, Sharp JM (2006) How organisational culture impacts on the implementation of ISO 14001: 1996—a UK multiple-case view. J Manuf Technol Manage 17 (1):89–103
Balzarova MA, Castka P (2008) Underlying mechanisms in the maintenance of ISO 14001 environmental management system. J Clean Prod 16:1949–1957
Beattie KR (1999) Implementing ISO 9000: a study of its benefits among Australian organizations. Total Qual Manage 10:95–106
Bénézech D, Lambert G, Lanoux B, Lerch C, Loos-Baroin J (2001) Completion of knowledge codification: an illustration through the ISO 9000 standards implementation process. Res Policy 30:1395–1407
Boiral O (2003) ISO 9000: outside the iron cage. Org Sci 14(6):720–737
Boiral O (2007) Corporate greening through ISO 14001: a rational myth? Org Sci 18(1):127–146
Boiral O (2011) Managing with ISO systems: lessons from practice. Long Range Plan 44:197–220
Boiral O (2012) ISO certificates as organizational degrees? Beyond the rational myths of the certification process. Org Stud 33:633–654
Boiral O, Guillaumie L, Heras Saizarbitoria I, Tayo Tene CV (2017) Adoption and outcomes of ISO 14001: a systematic review. Int J Manag Rev
Boiral O, Roy M (2007) ISO 9000: integration rationales and organizational impacts. Int J Oper Prod Manage 27:226–247
Burdick D (2001) American and European ISO 14001 accreditation requirements and their influences on registrar practice and environmental performance. Corp Environ Strateg 8 (1):65–74
Capehart BL, Turner WC, Kennedy WJ (2012) Guide to energy management. Energy Manag
Castka P, Corbett CJ (2015) Management systems standards: diffusion, impact and governance of ISO 9000, ISO 14000, and other management standards. Found Trends Technol Inf Oper Manage 7(3–4):161–379
Castka P, Prajogo D, Sohal A, Yeung ACL (2015) Understanding firms selection of their ISO 9000 third-party certifiers. Int J Prod Econ 162:125–133
Christmann P, Taylor G (2011) Standards: determinants of symbolic. Int Bus 37
Chung SS, Fryxell GE, Lo CW (2005) Corporate environmental policy statements in mainland China: to what extent do they conform to ISO 14000 documentation? Environ Manage 35 (4):468–482
Darnall N, Kim Y (2012) Which types of environmental management systems are related to greater environmental improvements? Public Adm Rev 72:351–365
Dickins DE, Higgs JL, Skantz TR (2008) Estimating audit fees post-SOX. Curr Issues Audit 2(1): A9–A18
Dogui K, Boiral O, Heras-Saizarbitoria I (2014) Audit fees and auditor independence: the case of ISO 14001 certification. Int J Audit 18:14–26
Du Plessis W (2015) Energy efficiency and the law: a multidisciplinary approach. S Afr J Sci 111:1–8
Dzene I, Polikarpova I, Zogla L, Rosa M (2015) Application of ISO 50001 for implementation of sustainable energy action plans. Energy Procedia 72:111–118
Epstein MJ, Roy MJ (1997) Using ISO 14000 for improved organizational learning and environmental management. Environ Q Manage 7(1):21–30
Fernández-Muñiz B, Montes-Peón JM, Vázquez-Ordás CJ (2012) Occupational risk management under the OHSAS 18001 standard: analysis of perceptions and attitudes of certified firms. J Clean Prod 24:36–47
Fryxell GE, Chung SS, Lo CWH (2004) Does the selection of ISO 14001 registrars matter? Registrar reputation and environmental policy statements in China. J Environ Manage 71:45–57

Fuentes CM, Benavant FB, Moreno MAE, Cruz TFG, del Val MP (2003) ISO 9000 based quality assurance approaches and their relationship with strategic analysis, Int J Qual Reliab Manag 20 (6/7):664–690
Giacone E, Mancò S (2012) Energy efficiency measurement in industrial processes. Energy 38:331–345
González-Benito J, González-Benito Ó (2005) Environmental proactivity and business performance: an empirical analysis. Omega 33(1):1–15
Gopalakrishnan B, Ramamoorthy K, Crowe E (2014) A structured approach for facilitating the implementation of ISO 50001 standard in the manufacturing sector. Sustain Energy Technol Assess 7:154–165
Gordić D, Babić M, Jovičić N, Šušteršič V, Končalović D, Jelić D (2010) Development of energy management system—case study of Serbian car manufacturer. Energy Convers Manage 51:2783–2790
Gotel DG (1989) The application of monitoring & targeting to energy management. HMSO, London
Granerud RL, Rocha RS (2011) Organisational learning and continuous improvement of health and safety in certified manufacturers. Saf Sci 49(7):1030–1039
Gyani G (2008) Effectiveness of QMS certification process. Total Qual Manage 19:263–279
Harris P (1989) Energy monitoring and target setting using CUSUM. Tech Pub, Cheriton
Heras-Saizarbitoria I, Boiral O (2013) ISO 9001 and ISO 14001: towards a research agenda on management system standards. Int J Manage Rev 15(1):47–65
Heras-Saizarbitoria I, Dogui K, Boiral O (2013) Shedding light on ISO 14001 certification audits. J Clean Prod 51:88–98
Hobbs RJ, Harris JA (2001) Restoration ecology: repairing the earth's ecosystems in the new millennium. Restor Ecol 9(2):239–246
Hooke JH, Landry BJ, Hart D (2003) Energy management information systems. Achieving improved energy efficiency. A handbook for managers, engineers and operational staff
International Energy Agency (2014) Key world energy statistics 2014. International Energy Agency, Paris
Imel MR, Gastesi R, Stone R (2015) Monroe County, Florida A case study in sustainable energy management. Energy Eng 8595:37–41
International Organization for Standardization (2011) ISO 50001:2011 energy management systems—requirements with guidance for use. ISO, Geneva
International Organization for Standardization (ISO) (2014) The ISO survey of management systems standards. ISO, Geneva
International Organization for Standardization (2015a) The ISO survey of management system standard certifications: executive summary
International Organization for Standardization (2015b) ISO 9001:2015 quality management systems—requirements with guidance for use. ISO, Geneva
International Organization for Standardization (2015c) ISO 14001:2015 environmental management systems—requirements with guidance for use. ISO, Geneva
Introna V, Cesarotti V, Benedetti M, Biagiotti S, Rotunno R (2014) Energy management maturity model: an organizational tool to foster the continuous reduction of energy consumption in companies. J Clean Prod 83:108–117
Jacobs BW, Singhal VR, Subramanian R (2010) An empirical investigation of environmental performance and the market value of the firm. J Oper Manage 28(5):430–441
Jiang RJ, Bansal P (2003) Seeing the need for ISO 14001. J Manage Stud 40(4):1047–1067
Johnson H, Johansson M, Andersson K, Södahl B (2013) Will the ship energy efficiency management plan reduce $CO_2$ emissions? A comparison with ISO 50001 and the ISM code. Marit Policy Manage 40:177–190
Jovanović B, Filipović J (2016) ISO 50001 standard-based energy management maturity model–proposal and validation in industry. J Clean Prod 112:2744–2755
Kanji GK, Asher M (1996) 100 methods for total quality management. Sage, London

Karapetrovic S, Casadesús M (2009) Implementing environmental with other standardized management systems: scope, sequence, time and integration. J Clean Prod 17:533–540
Karcher P, Jochem R (2015) Success factors and organizational approaches for the implementation of energy management systems according to ISO 50001. TQM J 27:361–381
Kaye M, Anderson R (1999) Continuous improvement: the ten essential criteria. Int J Qual Reliab Manage 16(5):485–509
Kaye MM, Dyason MD (1995) The fifth era. TQM Mag 7(1):33–37
Kaygusuz K (2007) Energy for sustainable development: key issues and challenges. Energy Sour B Econ Plann Policy 2:73–83
Kaygusuz K, Yüksek Ö, Sari A (2007) Renewable energy sources in the european union: markets and capacity. Energy Sour B Econ Plan Policy 2:19–29
King A, Lenox MJ (2009) Lean and green? An empirical examination of the relationship between lean production and environmental performance. Prod Oper Manage 10:244–256
Lafuente E, Bayo-Moriones A, Garcia-Cestona M (2010) ISO-9000 certification and ownership structure: effects upon firm performance. Br J Manage 21:649–665
Lambert DM, Stock JR (1979) The corporate energy policy: a management planning perspective. Long Range Plan 12(2):45–51
Lammers N, Kissock K, Abels B, Sever F (2011) Measuring progress with normalized energy intensity. SAE Int J Mater Manuf. doi:10.4271/2011-01-0320
Laskurain I, Heras-Saizarbitoria I, Casadesús M (2015) Fostering renewable energy sources by standards for environmental and energy management. Renew Sustain Energy Rev 50:1148–1156
Laskurain I, Heras-Saizarbitoria I, Casadesús M (forthcoming) Do energy management systems add value to firms with environmental management systems? Environ Eng Manage J
Lee J, Yuvamitra K, Guiberteau K, Kozman T (2014) Six-sigma approach to energy management planning. Strateg Plan Energy Environ 33:23–40
Leventis S, Weetman P, Caramanis C (2005) Determinants of audit report lag: some evidence from the Athens Stock Exchange. Int J Audit 9(1):45–58
Levine DI, Toffel MW (2010) Quality management and job quality: how the ISO 9001 standard for quality management systems affects employees and employers. Manage Sci 56(6):978–996
Lo VH, Sculli D (1998) Role of the management representative in quality management projects. Int J Contin Eng Educ Life Long Learn 8(1–2):76–85
Lo CK, Yeung ACL, Cheng TCE (2009) ISO 9000 and supply chain efficiency: Empirical evidence on inventory and account receivable days. Int J Prod Econ 118:367–374
Majernik M, Bosak M, Stofova L, Szaryszova P (2015) Innovative model of integrated energy management in companies. Qual Innov Prosper 19:22–32
Martínez-Costa M, Martínez-Lorente AR, Choi TY (2008) Simultaneous consideration of TQM and ISO 9000 on performance and motivation: an empirical study of Spanish companies. Int J Prod Econ 113:23–39
McClelland L, Cook SW (1980) Energy conservation in university buildings encouraging and evaluating reductions in occupants' electricity use. Eval Rev 4(1):119–133
Montalván SM, Chang JT (2006) Is there a market payoff for being green at the Lima Stock Exchange? In: Sustainability accounting and reporting, pp 251–280
Morrow D, Rondinelli D (2002) Adopting corporate environmental management systems: motivations and results of ISO 14001 and EMAS certification. Eur Manage J 20(2):159–171
Naveh E, Marcus A (2005) Achieving competitive advantage through implementing a replicable management standard: installing and using ISO 9000. J Oper Manage 24:1–26
Nicolau JL, Sellers R (2002) The stock market's reaction to quality certification: empirical evidence from Spain. Eur J Oper Res 142(3):632–641
O'Callaghan PW, Probert SD (1977) Energy management. Appl Energy 3:127–138
Palm J (2009) Placing barriers to industrial energy efficiency in a social context: a discussion of lifestyle categorisation. Energy Effic 2(3):263–270
Peter M (1995) ISO 9000 on the road to total quality. Manage Account Mag 69:21–26

Poksinska B, Dahlgaard JJ, Eklund JE (2003) Implementing ISO 14000 in Sweden: motives, benefits and comparisons with ISO 9000. Int J Qual Reliab Manage 20:585–606

Poksinska B, Jörn Dahlgaard J, Eklund JA (2006) From compliance to value-added auditing—experiences from Swedish ISO 9001: 2000 certified organisations. Total Qual Manage Bus Excell 17(7):879–892

Poksinska B (2007) Does standardization have a negative impact on working conditions? Hum Factors Ergon Manuf 17:383–394

Potoski M, Prakash A (2005) Green clubs and voluntary governance: ISO 14001 and firms' regulatory compliance. Am J Pol Sci 49:235–248

Power D, Terziovski M (2007) Quality audit roles and skills: perceptions of non-financial auditors and their clients. J Oper Manage 25(1):126–147

Prajogo D, Tang AKY, Lai KH (2012) Do firms get what they want from ISO 14001 adoption? An Australian perspective. J Clean Prod 33:117–126

Riche J-P (2013) Leanergy: how lean manufacturing can improve energy efficiency. Chim Int J Chem 67:700–702

Rodgers R, Hunter JE (1991) Impact of management by objectives on organizational productivity. J Appl Psychol 76(2):322

Rondinelli D, Vastag G (2000) Panacea, common sense, or just a label? The value of ISO 14001 environmental management systems. Eur Manage J 18(5):499–510

Russo MV (2009) Explaining the impact of ISO 14001 on emission performance: a dynamic capabilities perspective on process and learning. Bus Strateg Environ 18(5):307–319

Sambasivan M, Fei NY (2008) Evaluation of critical success factors of implementation of ISO 14001 using analytic hierarchy process (AHP): a case study from Malaysia. J Clean Prod 16 (13):1424–1433

Sammalisto K, Brorson T (2008) Training and communication in the implementation of environmental management systems (ISO 14001): a case study at the University of Gävle, Sweden. J Clean Prod 16(3):299–309

Santos G, Mendes F, Barbosa J (2011) Certification and integration of management systems: the experience of Portuguese small and medium enterprises. J Clean Prod 19:1965–1974

Siciliano G, de los Reyes P, Kramer C, Björkman T, Dahlgren M, Noda F, Ogawa J, Yamashita Y (2015) Models for driving energy efficiency nationally using energy management. Strateg Plan Energy Environ 35:48–79

Singh PJ, Feng M, Smith A (2006) ISO 9000 series of standards: comparison of manufacturing and service organisations. Int J Qual Reliab Manage 23:122–142

Sovacool BK (2009) Rejecting renewables: the socio-technical impediments to renewable electricity in the United States. Energy Policy 37:4500–4513

Stahl MJ, Grigsby DW (1997) Strategic management; total quality and global competition. Blackwell, Oxford

Stenqvist C, Nilsson LJ (2012) Energy efficiency in energy-intensive industries—an evaluation of the Swedish voluntary agreement PFE. Energ Effic 5:225–241

Switzer J, Ehrenfeld J, Milledge V (1999) ISO 14001 and environmental goal setting: promises kept. Environ Qual Manage 9(2):1–24

Tarasovskii VG, Petukhov IS (2014) Use of analytical-information systems to manage energy efficiency at metallurgical plants. Metallurgist 58(3–4):167–172

Terziovski M, Power D (2007) Increasing ISO 9000 certification benefits: a continuous improvement approach. Int J Qual Reliab Manage 24:141–163

Testa F, Vigolo V (2015) Sustainability through energy efficiency: an Italian perspective. Sinergie Ital J Manage 33:93–111

Therkelsen P, McKane A (2013) Implementation and rejection of industrial steam system energy efficiency measures. Energy pol 57:318–328

Therkelsen P, Mckane A, Scheihing P (2013) Assessing the costs and benefits of the superior energy performance program 14

Van der Wiele T, Van Iwaarden J, Brown A, Steimle U, Zink KJ (2009) An international comparison of the perceptions about the revised ISO 9000 quality system standards. Total Qual Manag 20(4):393–408

Westphal JD, Gulati R, Shortell SM (1997) Customization or conformity? An institutional and network perspective on the content and consequences of TQM adoption. Admin Sci Q 366–394

Worledge DH (1992) Role of human performance in energy systems management. Annu Rev Energy Environ 17(1):285–300

Wulandari M, Laskurain I, Fa MC, Heras-Saizarbitoria I (2015) Early adoption of ISO 50001 standard: an empirical study. In: Sustainable operations management. Springer, Berlin, pp 183–202

Yin H, Schmeidler PJ (2009) Why do standardized ISO 14001 environmental management systems lead to heterogeneous environmental outcomes? Bus Strategy Environ 18(7):469–486

Yiu L, Saner R (2005) Does it pay to train? ISO 10015 assures the quality and return on investment of training. ISO Manage Syst 5(2):9–13

Zobel T (2008) Characterization of the environmental policy implementation in an EMS context: a multiple case study in Sweden. J Clean Prod 16:37–50

# Conflicts Arising in the Generation Process of the ISO 45001 Standard

**Iñaki Heras-Saizarbitoria, Ander Ibarloza and Alberto Díaz de Junguitu**

**Abstract** The process involved in the design, acceptance and launching of the ISO 45001 standard is turning out to be long and tortuous. It has been confirmed that, like its predecessors in the environmental (ISO 14001) and corporate social responsibility (ISO 26000) spheres of activity, this may prove to be conflictive as it deals with substantive social aspects such as those referring to labour issues. Disagreements can easily arise between stakeholders, and although some approaches may not show evidence of this, occupational health and safety is a complex area that is replete with multidisciplinary components that influence the daily tasks performed by employees—multidisciplinary components that are so deep-seated and controversial as cultural, political and ethical matters. The design and launching process for standards such as ISO 45001 has been under-researched in scholarly literature that focuses on the study of meta-standards. Priority is given to those of a technical nature or those that simply appear to rehash previous literature with regard to the phenomenon itself subject to study. In order to fill this gap, this contribution based on a work in progress aims at shedding light on the process involved in generating the ISO 45001. Extensive field work has been designed for such purpose that was currently underway at the time this contribution was completed, given that the ISO 45001 design and launching process has not yet concluded. Preliminary results obtained from the work evidence two types of main discrepancy—on the

I. Heras-Saizarbitoria (✉)
Department of Management, Faculty of Economics and Business,
University of the Basque Country UPV-EHU, Plaza Oñati 1,
20018 San Sebastian, Spain
e-mail: iheras@ehu.es

A. Ibarloza
Department of Financial Economics I, Faculty of Economics and Business,
University of the Basque Country UPV-EHU, Bilbao, Spain
e-mail: ander.ibarloza@ehu.es

A.D. de Junguitu
Department of Applied Economics I, Faculty of Economics and Business,
University of the Basque Country UPV-EHU, Plaza Oñati 1,
20018 San Sebastian, Spain
e-mail: alberto.dj@ehu.es

 
I. Heras-Saizarbitoria (ed.), *ISO 9001, ISO 14001, and New Management Standards*,
Measuring Operations Performance, https://doi.org/10.1007/978-3-319-65675-5_10

one hand, discrepancies in terms of form and specific content and, on the other, underlying discrepancies such as those that raise doubts about the raison d'être of the standard itself.

**Keywords** ISO 45001 · Occupational health and safety · Management system standards · Meta-standards · Production of standards

## 1 Introduction

The ISO management system standards, also called ISO meta-standards, are voluntary management standards to formalize, systematize, and legitimize a very diverse set of managerial activities or tasks (Boiral and Heras-Saizarbitoria 2015). ISO meta-standards encompass a very wide range of aspects of business activity, such as quality management (e.g. ISO 9001) and environmental management (e.g. ISO 14001). These standards tend to use a quite similar methodology with regard to their creation, structure, implementation process and monitoring or auditing by a third party.

The generation and dissemination process for ISO standards has very diverse implications. Thus, one should refer to the implication of this phenomenon with that of business self-regulation which is sidestepped on too many occasions. As highlighted by authors such as Brunsson and Jacobsson, Christmann and Taylor and O'Rourke, the inability or unwillingness on the part of the governments of countries to pursue public regulation in certain areas of activity such as environmental issues or the one we are focusing on here—the rights and duties of workers within the transnational milieu—has intensified interest in self-regulation on the part of companies, and this is something that also greatly affects the management standard phenomenon (Brunsson and Jacobsson 2000; Christmann and Taylor 2001; O'Rourke 2006). Within this context, Potoski and Prakash (2005) consider that voluntary programmes emerge to a great extent owing to the high cost that sustaining monitoring and surveillance costs attached to the mandate and control approach entails for administration and as a complement to the latter. In these authors' opinion, voluntary regulation will prove more suitable insofar as it fits in more appropriately with existing legislation, given that cooperation between companies and regulators tends to be more effective than conflict. Moreover, in their opinion, the effectiveness of voluntary programmes requires certain context-related conditions such as a developed economic system, the existence of brand identities both of product and company, and rigorous regulators. To this end, governments would have to ensure compatibility of voluntary programmes with existing institutions via a suitable combination of regulations and the reinforcement of voluntary practices. In this respect, these authors consider that voluntary programmes may be understood as an attempt to build mutual trust between companies and regulators by institutionalising the environmental commitment and commitments of another nature on the part of companies in cooperation with governments.

For Boiral (2004), self-regulation represents a new way of governing institutions in accordance with current trends in public policies, which are marked by a logic regarding commitment to and conciliation with different demands rather than by a regulatory control and confrontation-based approach. In his opinion, these two types of logic—one emergent and the other interventionist—are not mutually exclusive but, in fact, genuinely complementary. This author warns us, however, that the use of standards which do not derive from regulations deriving from public initiative but rather that emanate from company decisions and agreements between market forces, cannot under any circumstances lead to a loss of legitimacy or the need for state intervention in this domain. For this reason, some authors characterise voluntary regulation as a hybrid set of rules in which both public and private institutions are closely involved in creating and disseminating principles, standards, programmes and systems (Clapp 1998; Mendel 2002). Hence, King et al. (2005) consider certifiable ISO standards in the field of the environment or occupational health and safety (OHS), one of the most widespread and legitimate self-regulation systems in the world, to be decentralised private institutions. This is because they are characterised, apart from by their voluntary adoption, by the fact that the supposed persons in charge of recognising participants and sanctioning the rest tend to be mainly diverse private agents rather than a central public authority.

However, as has been stated, this type of implication is sidestepped in many specialist approaches or techniques in such a way as to make it difficult to advance on many occasions in terms of academic, rigorous knowledge of these types of specific phenomena such as ISO standards. This is the more specific case with studies that analyse the launching of new standards—a line of research that faces two problems, among others. On the one hand, most analyses that are undertaken to analyse these self-regulation elements from a general standpoint fail to do so in depth with regard to the tools themselves. On the other, in the case of studies that focus on analysis of new standards or on their emergence process, such analysis tends to focus more on superficial issues highlighted by stakeholders as promoters of these standards or certification bodies (who have an obvious business interest in their dissemination). However, some works have been published in scholarly literature which prove to be the exception. This is the case with the works by Tammhallstrôm (e.g. Tammhallstrôm 1996) and their followers (e.g. Boström and Hallström 2013), aimed at analysing the production process of international management standards. In the case of the ISO 14001 standard, there were very few works that analysed the launching of the main global environmental standard (among which mention attention should without doubt be drawn to that by Haufler 1999), a process that proved very controversial. In the case of the ISO 26000 standard, given the controversy that accompanied its emergence process, it was in fact the focus of attention on the part of academics (e.g. Schmiedeknecht 2008; Balzarova and Castka 2012; Bres 2013).

In the case of the emergence of ISO 45001—a new ISO project to enact the meta-standard in the field of OHS—to the best of our knowledge no previous scholarly work has been published on the standard's gestation process. From the practitioner standpoint, attention should be drawn to the more or less widespread reports published by some stakeholders, among which priority is given to works that focus on the standard's alleged virtues. At the same time, works have also been published that are critical of the standard creation process and of its very raison d'être on the part of other agents, such as some representatives of trade union organisations (e.g. Robertson 2016; Torres 2017). Taking into account this gap in the literature, the aim of the present study is to shed light on the process of generation of the ISO 45001 standard, in order to analyze the conflicts evidenced in the mentioned process. For that purpose, an extensive qualitative exploratory and inductive empirical study was designed which involved intensive desk analysis (e.g. aimed at collecting the perspective of different stakeholders involved in the process), four semi-structured in-depth interviews with members of the Spanish ISO/PC 283 National Mirror Committee, an intense exchange of information through e-mails, telephone and Skype talks with very diverse stakeholders of the process and an in-depth analysis of internal documentation such as the minutes and the position papers of the ISO/PC 283 and the ISO/PC 283 National Mirror Committees. The field-work started in November 2016 and was still in progress at the time this chapter was finished in May 2017. The remainder of this chapter is arranged as follows. Following this introduction, a short review of the literature that deals with the standardization in the field of OHS. In the third section the preliminary results of the research are summarized. Lastly, the preliminary conclusions are presented and the avenues for future research are shortly discussed.

## 2 ISO 45001 and Standardisation in the Field of OHS

The process involved in launching an ISO standard in the field of OHS in reality constitutes a second attempt for this launching. The facts go back a long way to a dispute between the ILO and the BSI, when in 1995 it was deemed not convenient to develop an ISO for occupational health and safety management (Rubio Romero 2002). Nonetheless, taking into account on the one hand the vacuum created by a lack of management standard in this field and, on the other, the success of the BS 5750 and BS 7750 standards, the BSI published a non-certifiable guide in May 1996: the BS 8800 guide to occupational health and safety management systems. There was an ever-increasing demand for a management standard that could be integrated into the ISO 9001 and ISO 14001 standards on the part of stakeholders with interests in the sector, and management standards, drafts, models and systems and guides started to proliferate (Redinger and Levine 1998). That was when two main approaches came to the fore. On the one hand, the ILO thought this proliferation of systems might give rise to major confusion as far as occupational health and safety management concepts were concerned (Plant and O'Reilly 2003). Thus, in 1997 guidelines governing occupational health and safety management systems started to be developed, which had a global, voluntary and non-certifiable design that could be integrated and appeared in 2001 in the form of ILO-OSH 2001 (ILO 2001; Knutsen 2006). On the other hand and in view of the growing demand for such systems, the BSI thought it convenient to create an ISO, and in 1998 put together a draft standard that could be certified as an ISO based on the structure of the BS 8800 guide that had not been approved. Thus, the OHSAS project group processed a global standard that was easy to integrate and compatible with the ISO 9001 and 14001, and so emerged the OHSAS 18001, which was published in 1999 with the clear intention of subsequently being transformed into ISO 18001. However, the International Confederation of Free Trade Unions (ICFTU), which merged in 2006 with the World Confederation of Labour (WCL) to form the International Trade Union Confederation (ITUC), opposed this, arguing that the standard might end up in the hands of large companies and that the appropriate space for dealing with occupational health and safety was the ILO (Rubio Romero 2002). Therefore, the ISO's attempts to develop a meta-standard for OHS met with strong opposition from international employer and union bodies, on the basis that this was a matter that should be dealt with through social dialogue (Robertson 2016).

The differences highlighted by the two approaches would appear evident. On the one hand, the ILA seemed to be better protecting workers' interests, including the participation of workers' representatives in management, and recommended free training for all company employees during working hours—hence, they garnered trade union support. On the other, the BSI appeared to offer a more pragmatic stance regarding company management, having a solid network at their disposal for disseminating and implementing ISO standards and their comfortable integration. As the years past and with the success in disseminating both ISO 9001 and the ISO 14001, the lack of an ISO standard governing health and safety that could be integrated into the package of an integrated, triple-certification system became increasingly obvious, especially for some stakeholders who had a clear business interest in the existence of such an ISO standard. Furthermore, there were over 40 OHSAS international standards adopted by different countries, alongside a great variety of standards, guidelines and audits based on OHSMS (occupational health and safety management systems) (Robson et al. 2007). Thus, in 2013 ISO accepted the proposal made by the OHSAS Project Group to start work on developing an international standard in this field. The ISO and the field of national standard bodies (NSBs) estimated that the future ISO 45001 would not offer any major surprises in terms of the concepts to be included which are mainly used in OHSAS 18001, nor in terms of structure, which would be based on a common, high-level structure to facilitate its alignment and compatibility with the ISO meta-standards. This structure would in turn be based on the SL Annex—a document approved by ISO in 2012 to standardise the structure of meta-standards and, in accordance with its promoters, improve the compatibility of these standards.

On this point, it should be mentioned that OHSAS 18001 was published for the first time in 1999, three years after ISO 9001 and which, from the outset, was compatible with the ISO 9001 and ISO 14001 standards, although the 2007 version improved on it and fitted in better with the environmental management ISOs. In this way, the ISO 9001, ISO 14001 and OHSAS 18001 started to be implemented as an integrated management systems package (Bernardo et al. 2009). As for its dissemination, 92,302 OHSAS 18001 certificates had been issued by 31st December 2011 according to a 2011 OHSAS survey. This represents an increase of 69% in the last two years, although there would remain a long way to go before the certificates issued by the ISO 14001 could be achieved and far longer for the ISO 9001 certificates.

Following this first attempt, ISO decided to set a new management standard in motion governing occupational hazards along similar lines to OHSAS 18001, in view of the pressure that had long been applied by many national standardisation bodies to do so. It has to be taken into account that voluntary international standards such as ISO are developed on a consensus basis and following a systematic process, via a 6-step process were ISO members and sector specific experts are involved (ISO 2015): proposal, preparatory, committee, enquiry, approval and publication. As underlined by Jones (2017) once the need for a standard is established (proposal), a committee to oversee its development is formed (for ISO 45001 this is ISO Project Committee 283 also known as ISO/PC 283); the committee establishes a Work Group (WG), made up of experts from member organizations (usually national standards bodies) and liaison members, which are organizations with specific interests and technical expertise (preparatory). The WG develops the first draft known as the Committee Draft CD1 (committee) and ballots it to all WG members to gain consensus that it is sufficiently developed to move to the next stage. If consensus is achieved (two-thirds majority approve and less than a quarter disapprove), then the standard can become a Draft International Standard (DIS) and can be made available for consultation to the wider public, this is the stage of enquiry (Jones 2017). In order to achieve consensus, the WG meets to resolve any comments made at the previous draft stage and need to be agreed by the majority of the WG and the parent committee—for ISO 45001 the ISO/PC 283 (Jones 2017). The complex process of consulting on the draft standard and resolving comments can happen several times on the journey to a suitable consensus, so it can take a considerable amount of time (Jones 2017), notably in complex cases such as the analyzed one. Once any issues with the DIS are resolved, the ISO standard moves to the Final Draft International Standard (FDIS) stage and is circulated to all the member bodies of the International Organization for Standardization. In order to receive the final approval (stage of approval) two-thirds of the members of the mentioned organization have to approve it and less than a quarter disapprove it and once approved, the final text of the ISO standard is sent for publication (stage of publication).

When ISO decided to go ahead with this complicated process, the first thing it did was to inform the International Labour Organisation (ILO) that it had drafted some guidelines for prevention management in 2001 and that this could be considered the only one of these features on a global level. The ILO accepted this invitation and in 2013 started work on drafting the standard numbered 45001 as *Health and Safety Management Systems*. In October 2013, the project's technical committee—ISO/PC 283—met in London to create the first draft copy of the ISO 45001 standard. Based on OHSAS 18001, the British standard was internationally recognised and approved for occupational health and safety. The first draft copy appeared in 2014 which placed emphasis on amendments with a view to producing a text on which there could be consensus. Over 18,000 amendments were presented, and the technical committee prepared an initial project which was put to the vote in 2015 among national committees. In that first vote, the draft version of the proposed standard obtained more votes in favour than against, although not in sufficient proportion for the standard to be made definitive (Torres 2017). Therefore, a process got underway again to redraft a new project, which was put to the vote in May 2016 when it also failed to gain sufficient support. The usual ISO standard development timescale is 3 years (Jones 2017), which gives adequate time to reach consensus and achieve a standard that has international support but in some cases, such as in the case of ISO 45001, the process can still take longer.

In its preamble, the draft version of the ISO 45001 standard justifies the need to enact a standard in the field of occupational health and safety. Thus, data published by the ILO is cited that refers to the figures linked to over 2.3 million deaths and over 300,000 non-fatal accidents. Bearing this in mind, it is pointed out in the draft version that in view of these figures, it would be difficult to refuse any international advances made in terms of prevention in health and safety at work. To this end, an occupational health and safety management system was proposed, as can be deduced from the literature cited in ISO/DIS 45001, which would group together two international lines in an attempt to reach consensus—the ILO and the BSI, as follows:

- On the one hand, we find the ILO-OSH 2001 (ILO 2001) guidelines regarding occupational health and safety management systems issued by the ILO, which include an occupational health and safety management system, international work standards (including those governing health and safety at work) and an audit matrix. The latter means that auditors will be able to measure an organisation's OSH performance, and compare it with the 16 items described in ILO-OSH 2001, with a view to identifying weaknesses, providing solutions for improvement, and continually improving OSH performance.
- On the other hand, there is the *Occupational Health and Safety Assessment Series* (OHSAS) 18001, the direct reference for which is the BS 8800 guide developed by the ISO collaborating body, the *British Standards Institution* (BSI), as was the case with the basic ISO 9000 and 14000. It should be pointed out that OHSAS 18001, which can be easily integrated and acts as a complement to the ISO 9001 and 14001 standards, enjoys more support than the ILO-OSH in the practitioner field.

Thus, it was pointed out by its promoters that the purpose of ISO 45001 is to provide a reference framework in order to manage prevention of work-related deaths, injury and deterioration of health (ISO/DIS 45001). It is also stated that the aim of ISO 45001 is not to start from zero and to take advantage of the OHSAS's reputation and network; between the ILO and the BSI, it opts to pursue the approach set out by the BSI in facilitating adoption and establishing a suitable deadline for transition in order for companies certified according to OHSAS to be able to meet the requirements and obtain ISO 45001 certification. Indeed, it should be taken into account that ISO/DIS 45001 includes most of the requirements set out for the OHSAS 18001 standard.

## 3 Preliminary Results

The ISO 45001 gestation process is being characterised by the active involvement of different stakeholders who are highlighting the disparity in terms of standpoints. When the ISO/PC 281 committee was set up in 2013 with a view to obtaining the greatest possible consensus, the IOE (International Organisation of Employees), the ILO and the ITUC, among others, were invited to take part in the project and accepted, although the latter two did so cautiously. To get the standard-making process underway, the ILO issued the 2013 Memorandum of Understanding (MOU) between ISO and ILO that required ISO to "respect and support" the provisions made for ILO standards, not run counter to them. It was made clear in this document that "In case of conflict", the ILO standards would trump the ISO draft, which would need to be reworked accordingly (ILO 2017). For its part, the ITUC has stressed on several occasions to affiliates that ITUC involvement in this ISO process does not constitute any endorsement of the ISO setting these kinds of standard. They have on various occasions voiced their objections to the ISO entering into an area that they believe belongs to the ILO. Their involvement has been to protect the interest of workers in a standard that is likely to be widely used. This involvement is also important because of the participation of both the ILO and the International Organisation of Employers (lOE) in this process. In total there are 54 countries involved, 15 *observer* countries and a set of 17 *liaison members*.

Initially, the deadlines envisaged for the DIS project were by the end of 2014, the FDIS final draft by 2015 and the ISO standard by the end of 2016. However, these deadlines failed to be met owing to the above-mentioned disagreements. The first draft version appeared in 2014 in the amendment phase in order to obtain consensus about the text. Over 18,000 amendments were presented, and the technical committee drew up an initial project that was put to the vote in 2015 among the national committees. In that first voting, the proposed standard obtained more votes in favour than against, whereby a redrafting process for a new project once again got underway, which was also put to the vote in May of that year and also failed to garner sufficient support. In mid-May 2016, the ISO 45001 draft international standard was rejected. Of the 58 member bodies called on to vote, 28% voted against, including France, Germany, Italy, Spain, Australia, Canada and the United States and exceeded the 25% limit established (Eurogip 2016).

As underlined by Jones (2017), representative of the Institution of Occupational Safety and Health (IOSH), a British organization for health and safety professionals, in terms of comparison between ISO 45001 and OHSAS 18001, similarities include the use of the 'plan, do, check, act' model; the setting policies and objectives; and the methodology for conducting internal audits and review. Conversely, there are a number of differences including the following (Jones 2017):

- there are now ten sections; whereas OHSAS 18001 had four sections
- more focus on continual improvement and hierarchy of control
- more focus on risk management and ongoing assessment of risk and opportunities
- more focus on the need to demonstrate and understand compliance status at all times
- subsections on contractors, procurement and outsourcing
- more on use of performance indicators
- context of organization—such as its supply chain, local community and cultural, social, political, economic and governance environments
- leadership—top management need to take an active role e.g. strategic planning and leading organizational culture
- documented information—includes electronic and processed information e.g. use of smartphones or tablets.

The main practitioners of the field, such as the members of the standardization bodies, the members of the certification bodies and the consultants focus on these type of aspects related to the standard, frequently referred to as 'technical' aspects. For example, in an 'Expert Delphi' study carried out by IOSH in 2007 with the participation of some stakeholders (academic, business, regulators and health and safety professionals), the increased emphasis on leadership and context and greater focus on risk management contained in the new standard ISO 45001 when compared to the existing standard OHSAS 18001 was highlighted (Jones 2017).

Yet other stakeholders give a more negative perspective of the production process of the standard and of the ISO 45001 itself. Below are mentioned some of the main problems detected in the gestation phase of the standard, although the analysis is neither exhaustive nor definitive owing to space restrictions with the chapter and the fact that this is a process that has not yet concluded. Broadly speaking, two major sets of problems or confrontations have been detected in the course of the field work currently underway. On the one hand, there are those of a formal type and those with specific content, among which one of the most commonly-found problems is related to conceptual and terminological details. Firstly, one should refer to the problem with the definition of occupational risk itself, given that some stakeholders understand that the proposed text had serious shortcomings in this aspect, as it defined risk in such a way that it remained unclear and was a far cry from the concept of risk as set out by the European framework directive and the vast majority of national legal standards. Likewise, according to ISO/DIS 45001 requirements, the word *worker* is defined as a "person performing work or work-related activities under the control of the organisation". Yet the fact of the matter is that in the case of many organisations, workers are not limited to just employees and may also include contract workers and persons employed by other organisations. Another problem of this type is related to other elementary definitions of the term *workers' representative* that raised objections, among many others, by the ILO together with the definition of workplaces. As regards the latter, the rejected draft version of ISO 45001 proposed a very limited definition of workplace—associating it with a physical place owned by the company (e.g. a building, workshop or factory). This type of definition goes against the more common definition in modern prevention activity—an area in which the workplace is defined as any place where a worker does their job, whether on the premises owned by the company or in some other place (an aspect also highlighted by Torres 2017). As was pointed out to us by one of the people interviewed, "*the ISO* [45001] *as it is written does not consider work performed outside to be workplaces and this approach is incompatible with many legal precepts*". The concept of occupational disease was also subject to a confusing definition, in accordance with a trade union representative interviewed, given that the different ISO 45001 draft projects have not been made sufficiently clear in terms of what this refers to, confusing disease with accident.

Other problems go beyond merely conceptual considerations but also constitute specific allegations. Along these lines, we should refer to one of the criticisms made by the ILO, who published a paper on the draft standard in February, identifying areas of concern including the lack of any specification on the part of certified organisations to provide training and personal protective equipment free of charge (ILO 2017). The fact that workers' representatives have no guarantee of participation in the ISO 45001 field of application is also alluded to. On the other hand, who is in charge of the cost of personal protective equipment is unclear, when European legal regulations at least tend to state that this should be met by the company.

In terms of basic discrepancies, reference should be made to the fact that the trade unions have without doubt been the most critical stakeholders with the process and content of the first drafts of the standard. The ITUC has on the one hand criticised the fact that there has been ambiguity regarding the employer's obligation to meet all legal requirements and, on the other, with regard to workers' participation. The European Trade Union Confederation (ETUC) took a clearly opposing stance to the ISO 45001 project, stressing the fact that OHS is a fundamental worker's right and that the European Union has adopted 24 guidelines about OHS to provide a minimum level of protection in the workplace and also provides the basis for workers to enforce their rights. For this stakeholder, if ISO 45001 is adopted in the discussed form it would represent a risk for the advancement of the legal framework promoting good health and safety at work for the following reasons (ETUC 2016):

> 1. The proposed standard has the potential to undermine the EU OSH Framework: Non-compulsory standards should not deal with issues better regulated through legislation. (…)
>
> 2. Management System Standards (MSS) vs Legislation and binding standards: Our overall assessment is that the MSS in its current form promotes a process-driven approach relying on private bureaucracy and implicitly aimed at achieving certification. This encourages self-regulation endangering effective health and safety enforcement. OHS cannot be

reduced to a mere management system. The development of a certifiable MSS for OHS entails the risk that certification to ISO/DIS 45001 will be used as evidence of compliance with legal requirements.

3. Respect for ILO Conventions and fundamental workers' rights: The voluntary ISO/DIS 45001 standard is not the right place for establishing fundamental rights of workers regarding occupational health and safety. It does not respect and adequately reflect core principles of the ILO's International Labour Standards (ILS) in the field of OHS.

Other views of workers' representatives, such as some of those gathered in the course of field work, would seem to be more integrative and positive in terms of the contribution made by meta-standards: "*We are not against ISO 45001 or any other OHS management system—what we are trying to do is safeguard the rights of workers which has proved so difficult to regulate and remains so fragile, as has been apparent in the period of crisis. If implementation of ISO 45001 in companies helps to improve OHS, then this will mean major progress for everyone, although we'll have to try and establish certain guarantees to make sure that happens*".

Another interviewee stressed to us that the main basic problem, or at least the reason for the negative vote on the part of their National Mirror Committee, has been that the ISO/DIS 45001 technical standards have fewer requirements than those set out in legal standards governing the prevention of risks at work. They also pointed out to us that, in their opinion, the heart of the problem did not lie in disputes between companies and workers, but rather, focused on imbalances between *developed and emergent* countries (Hämäläinen 2009), in which the employment situation is more precarious. If more demanding technical standards were introduced, then many companies in *emergent* countries might not be able to comply with them either owing to the cost it would entail for the company or to the technical inability of their milieu—hence, the technical demands should be lowered in order to reach consensus. They also point out that "*those in charge of workers on the one hand see the danger that, instead of raising the level of technical requirements, the level of currently-established legal responsibilities might be lowered. On the other hand, these standards require greater technical demands in solidarity with workers who find themselves in a precarious employment situation, whether from emergent countries or otherwise*".

Although the critical view of trade unions about the basic problems has been highlighted in these preliminary results, it should be made clear that in terms of the negative vote by some National Mirror Committees, as in the case with Spain, the negative stances have not only been on the part of workers' representatives. Representatives of other interest groups such as those of small and medium-sized enterprises, those of employers' organisations and those of personal protective equipment suppliers also voted against.

Similarly, other relevant national and European multi-stakeholder boards or committees were also maintained a stance against the draft version of the ISO 45001 standard. For instance, the French Working Conditions Policy Board (COCT 2016)—in which trade unions, employers' organisations, the French Ministries of Labour and of Agriculture and the Directorate of Occupational Risks—voiced its opposition to the draft standard. This board underlined the fact that "the discussion process bearing on the ISO 45001 draft standard illustrates the need to set limits on the extent to which standards can intervene" (COCT 2016; p. 2). This committee also stressed the fact that the OHS standardisation phenomenon also gives rise to a major democratic problem due to three issues (COCT 2016; p. 3):

> - The guarantees associated with regulatory production and social dialogue are not found in standardisation procedures. In particular, the adoption of standards does not take on board the degree of legitimacy and representativeness of the stakeholders5;
>
> - The idea that a standard is exclusively voluntary is at odds with the reality of guidelines associated with supply chains, public purchasing and the legal reporting requirements of large companies, which may entail a de facto obligation for subcontractors to apply the standard, when access to standards is expensive. As for cross-cutting or social standards, such distribution is likely to jeopardise the application of legal rules relating to working conditions;
>
> - Certification resulting from certain standards would not provide sufficient assurances that legislation has been complied with in substantive terms. There are few guarantees that the principles provided for by the standard will be put into practice – fewer, at any rate, than in the case of regulations that are subject to scrutiny by the administration and judge.

The Advisory Committee on Safety and Health at Work (ACSH)—a tripartite body set up in 2003 that assists the European Commission in the field of OSH, made up of three full members per Member State, representing national governments, trade unions and employers' organisations—has also cast serious doubt and criticism on the draft ISO 45001 (ACSH 2016):

> The current draft of ISO 45001 will not be of real [stressed in the original] practical use for many employers, especially SMEs. Furthermore, OSH issues cannot be reduced to a management system, as they also have a strong human and social dimension, as well as legal and technical practicalities. Risk management is the responsibility of employers and in respect of the ISO standard 45001, currently being developed, onesingle model cannot be imposed on employers. Safeguarding occupational health and safety is a fundamental goal that must be built on the principles of social dialogue, where, depending on the different national industrial relations systems, can be set in national social partner agreements, and underpinned by European and national legislation. (…) The ISO international standard on an occupational health and safety management system has been developed by the standardization-certification systemfor its own sake. (…) Conformity to this standard does not prove compliance with the legislation or even the existence of efficient measuresof prevention at the workplace. It is only a pure management standard. If applied within a context of strong commitment to occupational health and safety, good outcomes, social dialogue and consultation of workers (in accordance with national industrial relations systems), as well as being underpinned by effective regulations which impose minimum requirements, this standard can be useful. If not, it will give the illusion of safety at the workplace, and especially if there is an official certification of the standard, itcould in fact indirectly reduce the level of workplace safety. This standard can produce greater bureaucracy and burdens especially for SMEs.

## 4 Discussion and Preliminary Conclusions

Following the customary conceptual and terminological debates about the design and definition of any management standard—which on many occasions are viewed as evidence—there is in fact evidence to suggest that ISO 45001 design and consensus underlie other more relevant discrepancies. This means discrepancies such as the confusion that may arise from certification owing to the use given to this type of voluntary certification and doubts about their real added value. As mentioned to us by interviewees, it has been ascertained that companies, workers' representatives and consultants have divergent interests.

As in the case with other meta-standards, compliance with a series of procedures and formal principles related to the OHS management system does not certify compliance with a series of specific objectives, particular results and/or effective compliance with a certain regulation. However, the perception—which may not necessarily be the right one—that the company has obtained the certificate by having effectively complied with the measures attached to the labour regulations to which it is subject can be fostered using these outlines—and this is the main key to disagreements that have arisen in the preliminary field work that has been conducted.

Hugh Robertson, senior policy officer for health and safety at the UK Trade Union Congress (who wrote the briefing for the ETUI) probably gets to the root of the matter in his critical analysis (Robertson 2016) when he points out that ISO 45001 "*has the potential to undermine the EU OSH Framework. (…) Non-compulsory standards should not deal with issues better regulated through legislation*" (Robertson 2016; p. 4). This is a further example of the basic debate associated with the phenomenon of private international business regulation and soft law (Vogel 2008), and constitutes a debate between *soft law* and *hard law*, or the promotion of self-regulation or not. It would seem clear that attempting to design, agree on and disseminate voluntary meta-standards (soft law) related to areas regulated by mandatory standards (hard law) creates insecurity and uncertainty and leads to social conflict. Certification may cover up any preventive shortcomings as has already been ascertained in other areas such as environmental management and CSR. Robertson (2016) underlines that previous initiatives of the field of OHS such as the 'Voluntary Protection Programs' in the US "*have shown that regulators have been willing to allow a level of self-regulation, often with disastrous results*" (Robertson 2016; p. 3). As pointed out by Hauert and Graz (2011) "*the growing influence of private labour standards has reinforced the need to address enduring question about their impact on traditional state-based labour regulations. Are private labour standard a complement or a substitute to traditional labour regulations?*" (Hauert and Graz 2011).

In such a situation of uncertainty in which OHSAS 18001 tends to disappear and ISO 45001 fails to even get off the ground, companies where meta-standards have not been implemented or certified in the field of OHS are not being encouraged to plan their setting in motion. Looking to the future, the field work currently underway needs to be completed with new sources of information such as an analysis of the internal information available. For instance, from the detailed study of thousands of allegations and comments submitted by the different stakeholders regarding the different versions of the standard, conclusions of interest may be drawn along the lines proposed by Balzarova and Castka (2012) for ISO 26000, about the different standpoints and contributions made by each stakeholder to the ISO 45001 design and launching process. Likewise, the type established would have to be qualified, as for instance it is clear that certain conceptual needs conceal a fundamental discrepancy.

**Acknowledgements** This chapter is a result of a Research Group funded by the Basque Autonomous Government (Grupos de Investigación del Sistema Universitario Vasco; IT1073-16).

## References

ACSH (Advisory Committee on Health and Safety at Work) (2016) The Advisory Committee on Safety and Health at Work, Draft opinion of the ACSH on standardization, Doc. 779 –EN, Adopted on 01/06/2016 https://circabc.europa.eu/faces/jsp/extension/wai/navigation/container.jsp. Accessed 14 May 2017

Balzarova MA, Castka P (2012) Stakeholders' influence and contribution to social standards development: The case of multiple stakeholder approach to ISO 26000 development. J Bus Ethics 111(2):265–279

Bernardo M, Casadesus M, Karapetrovic S, Heras I (2009) How integrated are environmental, quality and other standardized management systems? An empirical study. J Clean Prod 17 (8):742–750

Boiral O (2004) Du développement durable aux normes ISO: peut-on certifier la "bonne conduite" des entreprises? Revue Internationale de Droit et Politique du Développement Durable 2 (2):91–114

Boiral O, Heras-Saizarbitoria I (2015) Management system standards, meta perspective. In: Dahlgaard-Park SM (ed) The Sage encyclopedia of quality and the service economy. Sage Publications, Thousand Oaks

Boström M, Hallström KT (2013) Global multi-stakeholder standard setters: how fragile are they? J Glob Ethics 9(1):93–110

Bres LP (2013) Membership strategies in pluralistic organizations: the case of ISO 26000. In: Academy of Management Proceedings, vol 2013, No 1, p 15343. Academy of Management

Brunsson N, Jacobsson B (eds) (2000) A world of standards. Oxford University Press, Oxford

Christmann P, Taylor G (2001) Globalization and the environment: determinants of firm self-regulation in China. J Int Bus Stud 32(3):439–458

Clapp J (1998) The privatization of global environmental governance: ISO 14000 and the developing world. Glob Gov 4:295–316

COCT (Conseil d'orientation des conditions de travail) (2016) Opinion of the COCT standing policy group on standardisation in the occupational health sector. http://travail-emploi.gouv.fr/IMG/pdf/20161103_avis_gpo_coct_normalisation-en.pdf. Accessed 14 May 2017

ETUC (2016) ETUC Resolution on ISO DIS 45001Occupational health and safety management systems. https://www.etuc.org/documents/etuc-resolution-iso-dis-45001occupational-health-and-safety-management-systems#.WRvkFNSLTDc. Accessed 14 May 2017

Eurogip (2016) Negative vote on the ISO 45001 draft standard. http://www.eurogip.fr/en/eurogip-infos-news?id=4275. Accessed 14 May 2017
Hämäläinen P (2009) The effect of globalization on occupational accidents.Saf Sci 47(6):733–742
Hauert C, Graz JC (2011) All standards are labor standards. Trade unions' participation in European standardisation bodies. In: Vladislav FV, Kai J (eds) Proceedings of the 16th EURAS Annual Standardisation Conference. Peer-reviewed, Verlag Mainz, pp 119–132
Haufler V (1999) Negotiating International Standards for environmental management systems: the ISO 14000 standards. UN Vision Project on Global Public Policy Networks, New York, NY
ILO I (2001) Guidelines on occupational safety and health management systems, ILO-OSH 2001. Switzerland
ILO I (2017) Further developments in relation to the International Organization for Standardization, including in the field of occupational safety and health (OSH). http://www.ilo.org/wcmsp5/groups/public/—ed_norm/—relconf/documents/meetingdocument/wcms_222296.pdf. Accessed 14 May 2017
ISO (2015) 6 steps to creating an ISO standard, June 19, 2015. http://isoupdate.com/resources/6-steps-to-creating-an-iso-standard/. Accessed 14 May 2017
Jones R (2017) ISO 45001 and the evolution of occupational health and safety management systems, IOSH paper. https://www.asse.org/assets/1/7/IOSH_Article_for_ISO_45001.pdf. Accessed 14 May 2017
King AA, Lenox MJ, Terlaak A (2005) The strategic use of decentralized institutions: exploring certification with the ISO 14001 management standard. Acad Manag J 48(6):1091–1108
Knutsen HM (2006) Geographies of informalization: conceptual dilemmas over social standards and informalization of labor. Enterprising Worlds 86:221–230
Mendel PJ (2002) International standardization and global governance: the spread of quality and environmental management standards. In: Hoffman AJ, Ventresca MJ (eds) Organizations, policy, and the natural environmental: institutional and strategic perspectives. Stanford University Press, California, pp 407–431
O'Rourke D (2006) Multi-stakeholder regulation: privatizing or socializing global labor standards? World Dev 34:899–918
Plant R, O'Reilly C (2003) The ILO's special action programme to combat forced labour. Int Labour Rev 142(1):73
Potoski M, Prakash A (2005) Green clubs and voluntary governance: ISO 14001 and firms' regulatory compliance. Am J Polit Sci 49(2):235–248
Redinger C, Levine S (1998) Development and evaluation of the Michigan occupational health and safety management system assessment instrument: a universal OHSMS performance measurement tool. Am Ind Hyg Assoc J 59(8):572–581
Robertson H (2016) A new ISO standard for occupational health and safety management systems: is this the right approach?. ETUI Policy Brief, European Economic, Employment and Social Policy, N° 3/2016, pp 1–5
Robson LS, Clarke JA, Cullen K, Bielecky A, Severin C, Bigelow PL, Mahood Q (2007) The effectiveness of occupational health and safety management system interventions: a systematic review. Saf Sci 45(3):329–353
Rubio Romero JC (2002) Gestión de la prevención de riesgos laborales; OHSAS 18001-directrices OIT para su integración con calidad y medio ambiente. Díaz de Santos, Madrid
Schmiedeknecht M (2008) ISO 26000-reflecting the process of a multi-stakeholder dialogue: an empirical study (No. 29/2008). KIeM Working Paper
Tammhallström K (1996) The production of management standards. Revue d'économie industrielle 75(1):61–76
Torres (2017) http://www.porexperiencia.com/articulo.asp?num=74&pag=20&titulo=El-debate-sobre-la-norma-ISO-45001-de-%93Sistemas-de-gestion-de-salud-y-seguridad%94
Vogel DJ (2008) Private global business regulation. Annu Rev Polit Sci 11:261–282

# A Supply Chain View on Certification Standards: Does Supply Chain Certification Improve Performance Outcomes?

**Frank Wiengarten, George Onofrei, Paul Humphreys and Brian Fynes**

**Abstract** This paper aims to report the results of an empirical study, examining the perceived performance implications of quality, environmental and occupational health and safety certifications (i.e., ISO 9001, ISO 14001, and OHSAS 18001) at the supply chain level considering certifications of both the buyer and supplier. To assess the perceived performance implication of certifications at the supply chain level we collected survey data in Ireland from manufacturing plants. Our results indicate that certifications at the supply chain level lead to higher perceived performance outcomes for environmental certifications (i.e., ISO 14001) and occupational health and safety certifications (i.e., OHSAS 18001). However, our results could not confirm our hypothesis that quality certification (i.e., 9001) at the supply chain level leads to higher perceived quality performance. This study represents one of the first attempts to assess the impact of certifications on its intended performance dimensions at the supply chain level. The research takes into consideration different certifications at both the buyer's and supplier's plant.

**Keywords** ISO 9001 · ISO 14001 · OHSAS 18001 · Performance management · Buyer and supplier certification

F. Wiengarten
Department of Operations, Innovation and Data Sciences, ESADE Business School, Sant Cugat, Spain
e-mail: frank.wiengarten@esade.edu

G. Onofrei (✉)
School of Business, Letterkenny Institute of Technology, Letterkenny, Ireland
e-mail: george.onofrei@lyit.ie

P. Humphreys
Department of Management and Leadership, Ulster University Business School, Newtownabbey, UK
e-mail: pk.humphreys@ulster.ac.uk

B. Fynes
School of Business, UCD Smurfit Graduate Business School, Blackrock, Ireland
e-mail: b.fynes@ucd.ie



I. Heras-Saizarbitoria (ed.), *ISO 9001, ISO 14001, and New Management Standards*, Measuring Operations Performance, https://doi.org/10.1007/978-3-319-65675-5_11

# 1 Introduction

Supply chain and purchasing managers are constantly seeking for verification of their suppliers' performance in multiple performance dimensions. Managers have frequently and successfully relied on external certifications such as ISO 9001, ISO 14001 and OHSAS 18001 in the supplier selection process. These certifications have helped companies in the sourcing process to at least preselect and qualify a pool of companies that have externally verified processes to ensure a minimum and consistent level of quality, environmental and occupational health and safety standards.

ISO 9001 specifies the requirements for a quality management system (QMS) that an organization must fulfil to demonstrate its ability to consistently provide products and services that meet applicable statutory and regulatory requirements (ISO 2009a). Through the ISO 14001 standard companies implement operational controls to manage environmental concerns that are aimed at improving the efficient use of natural resources (ISO 2009b). The OHSAS 18001 certification focuses on the area of occupational health and safety management systems (OHSMS), concentrating on the wellbeing of the workforce (Marshall et al. 2016). These three certification standards represent some of the most important standards for manufacturing and service industries and are widely implemented globally (ISO 2013; OHSAS Project Group 2011). The primary goal of these certifications is to achieve process compliance (Gray et al. 2015), which may result in enhancing product and process performance as well as improving a firm's reputation (Wiengarten et al. 2013). In terms of the supply chain, these certifications provide a signalling effect which reassures buyers that their suppliers are taking seriously issues related to quality, environment and health and safety (Wiengarten et al. 2013).

The performance implications of these certifications have been extensively explored in previous research (e.g., Melnyk et al. 2003; Corbett et al. 2005; Levine and Toffel 2010; Sharma 2005; Boiral and Henri 2012; Lo et al. 2014). Although negative or non-significant certification—performance relationships have been found the general consensus in the literature indicates that certifications improve the operational and financial performance of a company.

However, what has been explored to a much lesser extent is the extent of these certifications on performance at the supply chain level. Certifications have been applied by companies to signal their commitment and intention to enter new markets or to expand market share (Wiengarten et al. 2013). However, these certification benefits on performance have been predominantly assessed solely from a single (focal) company perspective, rather than considering the supply chain as the level of analysis. This paper investigates the impact of certifications at the supply chain level thorough comparing the performance implications of certifications. Specifically, in this paper we explore the following research questions: *To what extent is the buyer's performance (in terms of quality, environmental and OHS performance) affected by their key supplier's ISO 9001, ISO 14001 and OHSAS 18001 certification?*

To explore this research question we collected survey data in Ireland. Results indicate that in terms of ISO 14001 and OHSAS 18001 the environmental and occupational health and safety performance of a buyer is higher when their key supplier is ISO 14001 and OHSAS 18001 certified compared to cases where the supplier is not ISO 14001 or OHSAS 18001 certified. However, our results could not confirm this in terms of ISO 9001 and quality performance. The quality performance of the buyer was not higher when their key supplier was ISO 9001 certified as compared to cases when the key supply was not ISO 9001 certified. We discuss these results in terms of the implications for practice and theory.

## 2 Literature Review

### *2.1 Certifications and Performance*

In today's competitive environment, firms understand the importance of standardized management systems (Abad et al. 2014) and their customers expect them to seek certification of their products, services or operations. ISO 9001 is one of the most common and established standards across industries (Lo et al. 2013). Due to increased pressures from multiple stakeholders, such as customers, regulators, government and non-governmental organisations, sustainability certifications in environmental (ISO 14001) and social (OHSAS 18001) management have become globally diffused. As mentioned previously, extensive research has been carried out at the plant level performance effects of these certifications (Fernández-Muñiz et al. 2009; Qi et al. 2013; Singh et al. 2011; Gray et al. 2015). There are mixed views, concerning the effective contribution of these certifications to plant performance. Although, several researchers highlight the positive relationship between these certifications (ISO 9001, ISO 14001 and OHSAS 18001) and superior performance of certified companies, a number of studies suggest that these standards are either "tick box" exercises or process orientated with little impact on performance. It is important to note that the certification process does not guarantee a particular plant performance level, but rather prescribes the standardisation of the processes, that companies can use to enhance specific performance dimensions (Bernardo et al. 2009). A summary of research studies on the certifications (ISO 9001, ISO 14001 and OHSAS 18001) and their effect on performance, as well as the dimensions used to estimate firm performance, is presented in Table 1. The overall conclusion is that these certifications affect company performance in a variety of ways.

**Table 1** Impact of certifications on firm performance: summary of research

| Certification type | Research study | Does the certification affect performance? yes/no | Firm performance dimensions |
|---|---|---|---|
| ISO 9001 | Corbett et al. (2005) | Yes | Productivity improvements, market benefits and financial performance |
| | Morris (2006) | No | Financial performance |
| | Han et al. (2007) | No | Various business performance |
| | Benner and Veloso (2008) | Yes | Financial performance |
| | Dick et al. (2008) | No | Return on assets |
| | Lafuente et al. (2010) | Yes | ROA, labour productivity |
| | Levine and Toffel (2010) | Yes | Rates for sales, employment, payroll, and average annual earnings |
| | Srivastav (2011) | Yes | Organisational effectiveness and development |
| | Starke et al. (2012) | Yes | Sales revenue, cost of goods sold, asset turnover ratio |
| | Lo et al. (2013) | Yes | Operational and financial performance |
| | Huo et al. (2014) | Yes | Product and process flow management |
| | Chatzoglou et al. (2015) | Yes | Quality awareness, operations execution, market share, customer satisfaction and sales revenue |
| ISO 14001 | King et al. (2005) | No | Environmental performance |
| | Potoski and Prakash (2005) | Yes | Environmental performance |
| | Barla (2007) | No | Reduction of emissions |
| | Arimura et al. (2008) | Yes | Natural resource use, solid waste generation, and wastewater effluent |
| | Lam et al. (2011) | No | Environmental performance |
| | Nishitani et al. (2012) | Yes | Pollution emissions |
| | Blackman (2012) | No | Environmental performance |
| | Boiral and Henri (2012) | Yes | Environmental performance |
| | Zobel (2013) | No | Air emissions, water emissions, resource use, energy use, and waste |
| | Testa et al. (2014) | Yes | Pollution emission |
| | Campos et al. (2015) | Yes | Environmental performance |
| | Arimura et al. (2016) | Yes | Environmental performance |

(continued)

**Table 1** (continued)

| Certification type | Research study | Does the certification affect performance? yes/no | Firm performance dimensions |
|---|---|---|---|
| OHSAS 18001 | Chang and Liang (2009) | Yes | Safety index score |
| | Hohnen and Hasle (2011) | Yes | Safety standards |
| | Vinodkumar and Bhasi (2011) | Yes | Safety management and employee behaviour |
| | Choi et al. (2012) | Yes | Sales and ROA |
| | Fernández-Muñiz et al. (2012b) | Yes | Safety behaviour, employee satisfaction, and qualitative measures of business competitiveness |
| | Abad et al. (2013) | Yes | Safety and operating performance |
| | Lo et al. (2014) | Yes | Safety, sales growth, labour productivity, and profitability |

## 2.2 *Certifications at the Supply Chain Level*

Operating globally and the increased market pressures, force companies to focus on the overall supply chain impact of all triple bottom line dimensions: economic, environmental and social issues (Wilhelm et al. 2016). In the past two decades, supply chain management has been through a significant paradigm shift (Curkovic and Sroufe 2011); businesses recognised that they cannot compete as an individual entity, but rather using their supply chain (Lambert and Cooper 2000) and as a result, firms established cooperative and mutually beneficial relationships with their supply chain partners (Prajogo et al. 2012).

Rising external pressures from regulatory and societal bodies as well as customers has encouraged companies to focus on their suppliers in ensuring sustainable multi-tier supply chains (Wilhelm et al. 2016). This is particularly the case in manufacturing as they are perceived to have a greater social and environmental impact, relative to other sectors (Gualandris and Kalchschmidt 2016). For example, companies such as IKEA (Vachon and Klassen 2006), Sony Ericsson (Norrman and Jansson 2004), Mattel (Hora et al. 2011) and Zara (Burgen and Phillips 2011) have transferred these external sustainability pressures to their supply chain in order to prevent negative publicity. Since 1994, in the car manufacturing sector, Ford, General Motors and Chrysler now require all tool and equipment suppliers to obtain QS-9000 certification, which is an auto-industry specific version of ISO 9000 (Hwang et al. 2006). Hence, one approach that is used by many buyer organisations is to require that their suppliers should implement international management standards, such as ISO 9001, ISO 14001 and OHSAS 18001. However, the automotive example illustrated by Choi and Hong (2002), shows how difficult it is for companies to manage and exert control over the lower-tier suppliers. These firms, which are at the peripheral levels of the supply chain, tend to be small or medium enterprises with limited resources and less exposure to these external pressures. Therefore, the engagement with the first-tier supplier becomes paramount, in the successful implementation and management of these certifications in the supply chain (Grimm et al. 2014).

The literature on the impact of certifications at plant level is substantial, but less is known about the effect at the supply chain level (Prajogo et al. 2012; Wiengarten et al. 2013; Qi et al. 2013). In manufacturing industries, because they heavily rely on suppliers, the success of buyers is largely dependent on the performance outcomes of their suppliers (Krause et al. 2007). Therefore, further investigation is required to understand the impact of these certifications on the performance, at the supply chain level, specifically if the performance of the buyer is higher, when their direct supplier is certified.

### 2.2.1 Performance Implications of ISO 9001

Several studies (Dick et al. 2008; Levine and Toffel 2010; Psomas et al. 2013) claim that ISO 9001 certification improves both management practices and the performance of internal processes, and these improvements ultimately affect the financial performance. Levine and Toffel (2010) state that the sales turnover improvements are a result of customers interpreting the ISO 9001 certification as a signal of high-quality products or service.

Although ISO 9001 addresses both internal and external organisational processes, there is an abundance of studies that have only examined the implementation effect at plant level, however research at supply chain level is limited (Prajogo et al. 2012).

Singh et al. (2011) explored the use of the ISO 9001 standard as a tool to manage the organizational environment, specifically the relationships between process management (internal process, customer process and supplier process) and operating performance (as a measure of effectiveness). Their findings highlight that selected dimensions of process management have a higher impact on operating performance when considered collectively, rather than individually. A similar study by Prajogo et al. (2012), investigated three levels of ISO 9001 implementation (basic, advanced and supportive) on supply chain activities. Basic implementation included the basic quality practices and principles; advanced implementation reflected the philosophy of the standards; and supportive implementation represented the management system and supporting the basic and advanced implementation. The results show that both advanced and supportive implementations have a positive effect on the operational performance. The basic implementation displayed only an interactive effect, meaning that its influence depends on other aspects of the implementation process, which might explain why some companies do not gain benefits from the certification process. Hwang et al. (2006) provided empirical support for the buyer's increasing use of the ISO 9001 certification as an incentive for improving the supplier's quality. They found that buyers prefer the ISO 9001 certification process over the appraisal process (the traditional inspection regime) when dealing with suppliers.

ISO 9001 supplier certification means that processes within the company follow a certain quality management standard. Although it does not guarantee a certain level of quality performance, the certification supports the enhancement of supplier's quality objectives, which in turn may improve the buyer's quality performance. Thus, the following hypothesis is proposed:

H1: The quality performance of a buyer is higher when their key supplier is ISO 9001 certified compared to cases where the supplier is not ISO 9001 certified.

### 2.2.2 Performance Implications of ISO 14001

Environmental sustainability is a key and current concern for companies and recent research has explored this topic at the supply chain level (Gualandris and Kalchschmidt 2016; Wilhelm et al. 2016; Vanpoucke et al. 2014; Wiengarten et al. 2013). The volatile competitive environment has forced companies to rely on supply chains to source materials, components, manufacturing process and other services. This results in a greater dependency on suppliers and more attention is being given to their contribution to the overall environmental sustainability of the supply chain (Wilhelm et al. 2016). To respond to these pressures, many firms have invested in environmental management systems, such as ISO 14001 (Nawrocka et al. 2009). This standard is designed for companies to identify and establish the importance of their environmental performance. Through the implementation of ISO 14001, firms develop controls to manage their environmental impact and this enables them to improve the use of natural resources and reduce negative externalities (Boiral and Henri 2012). Similar to ISO 9001, the ISO 14001 standard does not guarantee environmental excellence or compliance, but rather describes a system that will help an organisation to achieve its environmental objectives. The expectation is that better environmental management will result in improved environmental performance (Curkovic and Sroufe 2011). The general finding of previous research on ISO 14001 is that certified companies report higher levels of environmental performance however non-significant results do also exist (de Vries et al. 2012).

The motivations for implementing an environmental management system vary (Campos et al. 2015; Wiengarten et al. 2013); these tend to be external factors such as customers (Pagell et al. 2010; Gualandris and Kalchschmidt 2014), ethical motivations (Bansal and Roth 2000; Lam et al. 2011), performance improvement (Montabon et al. 2007; Arimura et al. 2016), compliance with legislation (Blackman 2012; Gray et al. 2015), marketing and legitimating reasons (Boiral and Henri 2012; Qi et al. 2013) and other intangible benefits (i.e. communication, employee motivation) (Zutshi and Sohal 2004; Campos et al. 2015).

In this study we focus on the use of ISO 14001 implementation at the supply chain level (buyer and supplier) as an approach for environmental performance improvement. Firms that have implemented ISO 14001, need to address the environmental aspects with their suppliers in order to improve the overall supply chain performance (Nawrocka et al. 2009). The need for integration of the environmental aspects in purchasing activities has been highlighted (Handfield et al. 2005) and the ISO 14001 standard has been proposed as a management system for improving environmental performance in the procurement process (Koplin et al. 2007; Chen 2005; Lee et al. 2009). Darnall et al. (2008) investigated the effect of environmental certification on environmental supply chain activities and found that certified companies were more likely to improve their environmental performance than the non-certified companies. More recently, Arimura et al. (2011) reported that ISO 14001 certification promotes green supply chain management and certified companies are 40% more likely to measure their suppliers' environmental performance, with in excess of 50% of companies being more likely to demand that their suppliers follow specific environmental practices. Curkovic and Sroufe (2011) highlighted that through ISO 14001 certification firms can promote a sustainable supply chain strategy and this can be leveraged throughout the supply chain resulting in a competitive advantage. Furthermore, Chiarini (2012) analysed 18 companies that used ISO 14001 standard requirements for improving the supply chain environmental sustainability. The research used a five step approach through which the supplier obtained the status of "green partner". This approach allowed the buyer to define a set of environmental performance indicators and manage the improvements of the supplier. The overall effect of such collaboration was reported as strategic and innovative, however no direct link to the buyer environmental performance was assessed.

Given the global diffusion of ISO 14001 certified companies and the diverse motivations for implementing environmental supply chain initiatives, the following hypothesis is proposed:

H2: The environmental performance of a buyer is higher when their key supplier is ISO 14001 certified compared to cases where the supplier is not ISO 14001 certified.

### 2.2.3 Performance Implications of OHSAS 18001

OHSAS 18001 is a formal external certification formulated by international certifying bodies based on the British Standard 8800 (Qi et al. 2013) and it forms part of the occupational health and safety management systems (OHSMS). This standard is designed to help companies eliminate and minimise occupational health and safety risks to employees and other stakeholders, as part of their social responsibility (Fernández-Muñiz et al. 2012a, b; Bottani et al. 2009; Hohnen and Hasle 2011).

Abad et al. (2013) developed a taxonomy which suggests that there is a tendency among researchers to focus on evaluating the OHSAS 18001 benefits in organisations, and limited investigation has been done at the supply chain level. Gualandris and Kalchschmidt (2016) used a sample of Italian manufacturing firms to explore how environmental and social (i.e., health and safety indicators) performance of manufacturing companies can be improved through the development of sustainable supply chain management practices. Their findings show that the effect of external practices on manufacturing companies 'sustainability performance is fully mediated by the key supplier's sustainability'. This raises the importance of social sustainability standards such as OHSAS 18001, at the supply chain level and how it influences the overall social sustainability performance.

Another example, where health and safety issues are a core value of the supply chain is the Electricity Supply Board (ESB) Ireland. The supplier health and safety performance is paramount for ESB as it affects directly their services offered to the end-customer. The company uses a supplier charter to manage and control their health and safety performance (ESB 2016). The company expects that the suppliers provide a safe workplace for their employees in compliance with all applicable laws and regulations. When working at ESB (or ESB customer) sites, suppliers are expected to comply with all ESB Contractor Safety Regulations communicated to them as well as any site-specific safety instructions given to them by ESB.

In several manufacturing sectors such as, food and textile, researchers have highlighted the ethical and social responsibility that companies must consider when outsourcing their production worldwide and the impact that this has on the supply chain. For example, consumers have criticised apparel company NIKE in relation to the use of sweatshop labour issues at its suppliers (Maloni and Brown 2006) and it has had a negative impact on their corporate image. Lund-Thomsen and Lindgreen (2014) assert that the social performance of suppliers affects the overall supply chain performance and propose that companies develop cooperative initiatives for managing social responsibility in global value chains. Therefore, certifications such as OHSAS 18001, can provide a signalling effect that suppliers are emphasizing a high degree of health and safety performance (Qi et al. 2013).

Based on the limited number of research papers and the practical examples outlined above, our research asserts that OHSAS 18001 implementation at the supply chain level (buyer and supplier) can lead to superior performance and the following hypothesis is proposed:

H3: The occupational health and safety performance of a buyer is higher when their key supplier is OHSAS 18001 certified compared to cases where the supplier is not OHSAS 18001 certified.

## 3 Method

### *3.1 Data*

To test the combined impact of buyer and key supplier certifications on performance we collected data through a survey in Ireland. The level of analysis was the manufacturing plant and the respondents were plant managers. These key

**Table 2** Sample descriptives

| Industry | Frequency |
|---|---|
| Food and kindred products | 8 |
| Apparel and other finished products made from fabrics and similar materials | 1 |
| Chemicals and allied products | 3 |
| Rubber and miscellaneous plastics products | 7 |
| Primary metal industries | 4 |
| Fabricated metal products, except machinery and transportation equipment | 7 |
| Industrial and commercial machinery and computer equipment | 4 |
| Electronic and other electrical equipment and components, except computer equipment | 8 |
| Measuring, analysing, and controlling instruments; photographic, medical and optical goods; watches and clocks | 2 |
| Manufacture of motor vehicles, trailers and semi-trailers | 2 |
| Manufacture of other transport equipment | 3 |
| Stone, clay, glass, and concrete products | 2 |
| Miscellaneous manufacturing industries | 7 |
| Total | 59 |

| Size | Frequency | Buyers' Certifications | Frequency | Suppliers' Certifications | Frequency |
|---|---|---|---|---|---|
| 1–25 | 2 | ISO 9001 | 46 | | |
| 26–100 | 15 | ISO 14001 | 26 | | |
| 101–250 | 20 | OHSAS 18001 | 19 | | |
| 251–500 | 11 | | | | |
| 501–1000 | 8 | | | | |
| >1000 | 3 | | | | |

informants had the comprehensive knowledge related to the management and operations of the plant and they were advised to supplement this with input from other functions, where appropriate. The majority of the data was collected electronically via email. Other methods were used as well, such as telephone, mail and face-to-face interviews. Table 2 provides an overview of the dataset in terms of industry sector, company size and certification frequency. The data was collected at the end of 2014 and early 2015. The population chosen for this research was that of manufacturing plants in Ireland within the industry classification codes of SIC 27 and SIC 38 employing twenty or more people. The size of the population was established from a number of databases, including Kompass Ireland, the Industrial Development Authority and Enterprise Ireland. Given the SIC codes, 500 companies were identified and the response rate of just over 12% is satisfactory and in alignment with recent survey research in the operations management domain.

## 3.2 Measures

Operations performance was measured across the selected dimensions of quality, environmental and health and safety performance (Shin et al. 2000; Rosenzweig and Roth 2004; Pagell et al. 2014). Respondents were prompted to indicate their plant's performance compared to their major competitors. The scale ranged from one to seven where one means far worse, four means similar and seven far better (see Table 3).

Quality performance was measured using two items with regards to product performance and product conformance to customer specifications. Environmental performance was measured through prompting the respondents to indicate the extent to which your plant has performed from an environmental perspective during the past two years. The scale ranged from one to seven where one means not at all, four means to some extent seven to a great extent. Four items are used to represent the environmental performance dimension (see Table 3). Occupational health and safety performance was measured through the same scale as used for the environmental dimension. Again four items were used to represent this performance dimension, which are also listed in Table 3.

Buyer certification was measured through binary questions prompting the respondents to indicate "*Has your plant obtained any of the following certifications?*" (ISO 9001, ISO 14001, OHSAS 18001). Furthermore, we asked the buyer to indicate the certification status of their key suppliers through prompting

**Table 3** Construct measurement items

| Items | Mean | SD | Factor loading | Alpha values |
|---|---|---|---|---|
| *Quality performance* | 5.42 | .852 | | .719 |
| Product performance | | | .603 | |
| Product conformance to customer specifications | | | .923 | |
| *Environmental performance* | 4.58 | 1.344 | | .886 |
| We have reduced energy use in our facilities | | | .784 | |
| We have reduced water use in our facilities | | | .802 | |
| We have reduced waste at our facilities | | | .750 | |
| We have reduced emissions at of our facilities | | | .783 | |
| *Occupational health and safety performance* | 4.94 | 1.174 | | .953 |
| We have reduced the number of occupational-related accidents at our facilities | | | .844 | |
| We have reduced the number of occupational-related injuries at our facilities | | | .928 | |
| We have reduced occupational-related ill health at our facilities | | | .870 | |
| We have reduced the number of occupational-related insurance claims at our facilities | | | .831 | |

*"Considering your key supplier (in terms of strategic importance for your most important product line). To the best of your knowledge which of the following certifications do they possesses?"*.

Furthermore, we controlled for company size through number of employees. All latent variables are listed in Table 3.

## *3.3 Construct Validation*

We conducted exploratory factor analysis (EFA) to validate our measures and to confirm our proposed factor structure (using SPSS 20 for this and subsequent analyses). We decided on conducting EFA instead of confirmatory factor analysis (CFA) because of our relatively small sample size. Various scholars have called for having at least 100 (e.g., Kline 1979) or 150 (Hutcheson and Sofroniou 1999) cases to conduct CFA. Thus, we acknowledge that our measures are established and their factor structure has been confirmed in previous research. But recognize the limitations of our data set and thus conducted EFA. Specifically, we conducted principle axis factoring along with varimax rotation. As expected the EFA converged in a three-factor solution in terms of quality, environmental and occupational health and safety performance.

Results presented in Table 3 indicate relatively high factor loadings with the lowest value of .603. This can be interpreted as an initial indicator of the validity of our identified factor structure (Nunnally 1978). Furthermore, no cross-loadings were detected in our solution. The initial eigenvalue for the quality performance factor was 5.074, the environmental performance factor 1.679 and for the occupational health and safety performance factor 1.408. Resulting in a cumulative percentage of the initial values of 81.62%. The cumulative parentage of the rotation sums of squared loadings resulted in 73.57%. These are also further signs of construct validity.

Finally, Cronbach's alpha ($\alpha$) has been used to test for reliability. The Cronbach's alpha values listed in Table 3 are all above the commonly accepted level of .7, which indicates that reliability is satisfactory. Based on the above analyses, the validity and reliability of our scales were established.

Table 4 presents the Pearson correlation between the composite score of our explored factor structure. We will continue our analyses using the composite score to test our hypotheses.

## *3.4 Common Method Variance Bias*

We test for common method bias through conducting the Harman's one-factor test (Podsakoff et al. 2003). Thus we loaded all items on a non-specified factor in an un-rotated factor structure. The first factor accounts for 50.74% of variance, and the

**Table 4** Correlations

| Variables | (1) | (2) | (3) | (4) | (5) | (6) | (7) | (8) | (9) |
|---|---|---|---|---|---|---|---|---|---|
| Buyer's ISO 9001 certification (1) | 1 | | | | | | | | |
| Buyer's ISO 14001 certification (2) | .448** | 1 | | | | | | | |
| Buyer's OHSAS 18001 certification (3) | .346** | .772** | 1 | | | | | | |
| Supplier's ISO 9001 certification (4) | .632** | .289* | .221 | 1 | | | | | |
| Supplier's ISO 14001 certification (5) | .129 | .700** | .513** | .402** | 1 | | | | |
| Supplier's OHSAS 18001 certification (6) | .283* | .600** | .762** | .262* | .651** | 1 | | | |
| Quality performance (7) | −.486** | −.194 | −.116 | −.388** | −.230 | −.153 | 1 | | |
| Environmental performance (8) | −.005 | .331* | .328* | −.036 | .295* | .223 | .011 | 1 | |
| Occupational health and safety performance (9) | .186 | .422** | .312* | .255 | .425** | .326* | −.080 | .530** | 1 |
| Size (10) | −.052 | .051 | .091 | .101 | .318* | .348** | −.118 | .184 | .211 |

**Correlation is significant at the 0.01 level (2-tailed)
*Correlation is significant at the 0.05 level (2-tailed)

other items load on different factors. Thus, we expect that common method variance does not pose a threat to our data.

# 4 Results

To test our three hypotheses, we conducted a one-way analysis of variance (ANOVA). Specifically, we calculated three models to determine the effects of joint ISO 9001 (i.e., buyer and supplier are certified) on quality performance, joint ISO 14001 (i.e., buyer and supplier are certified) on environmental performance and joint OHSAS 18001 (i.e., buyer and supplier are certified) on occupational health and safety performance. Thus, the dependent variables were the three performance dimensions for each certification (i.e., quality, environmental, and occupational health and safety performance) and the fixed factors were the certification (i.e., ISO 9001, ISO 14001, OHSAS 18001). In Table 5 we report the performance differences for joint certification versus non-joint certification (i.e., supplier is not certified).

In Table 6 we report the results of the ANOVA test for our quality model. Results indicate a statistical difference between the two groups ($F = 16.200$; $p = .000$). However, contrary to our hypothesis one having both buyer and the supplier ISO 9001 certified leads to significantly lower quality (Quality Mean = 5.18) performance levels as if only the buyer would be certified (Quality Mean = 6.10).

**Table 5** Cross-Tab: certifications and performance

| Certification/performance | Joint ISO 9001/non-joint 9001 | Joint ISO 14001/non-joint 14001 | Joint OHSAS 18001/non-joint 18001 |
|---|---|---|---|
| Quality performance (mean; SD) | 5.18; .779/6.10; .686 | | |
| Environmental performance (mean; SD) | | 5.14; 1.20/4.19; 1.31 | |
| Occupational health and safety performance (mean; SD) | | | 5.59; 1.11/4.70; 1.11 |

**Table 6** ISO 9001 ANOVA results

| ISO 9001/quality performance | Sum of squares | Mean square | F | Sig. |
|---|---|---|---|---|
| Between groups | 9.289 | 9.289 | 16.200 | .000 |
| Within groups | 32.112 | .573 | | |
| Total | 41.401 | | | |

**Table 7** ISO 14001 ANOVA results

| ISO 14001/environmental performance | Sum of squares | Mean square | F | Sig. |
|---|---|---|---|---|
| Between groups | 12.930 | 12.930 | 8.022 | .006 |
| Within groups | 91.875 | 1.612 | | |
| Total | 104.805 | | | |

**Table 8** OHSAS 18001 ANOVA results

| OHSAS 18001/occupational health and safety performance | Sum of squares | Mean square | F | Sig. |
|---|---|---|---|---|
| Between groups | 9.122 | 9.122 | 7.339 | .009 |
| Within groups | 70.851 | 1.243 | | |
| Total | 79.972 | | | |

In Table 7 we report the results of the ANOVA test for our environmental model. Results indicate a statistical difference between the two groups ($F = 8.022$; $p = .006$). These results confirm our second hypothesis. Having both, buyers and the suppliers ISO 14001 certified leads to significantly higher environmental (Environmental Mean = 5.14) performance levels as if only the buyer would be certified (Environmental Mean = 4.19).

In Table 8 we report the results of the ANOVA test for our occupational health and safety model. Results indicate a statistical difference between the two groups ($F = 7.339$; $p = .009$). These results confirm our second hypothesis. Having both, buyers and the suppliers OHSAS 18001 certified leads to significantly higher occupational health and safety (Occupational Health and Safety Mean = 5.59) performance levels as if only the buyer would be certified (Occupational Health and Safety Mean = 4.70).

Thus, in concluding our empirical results we find that two of our three proposed hypotheses are supported (H2 and H3). We will discuss these results in the following sections in terms of their managerial and theoretical implications.

## 5 Discussion

The results confirm the hypotheses regarding environmental and occupational health and safety certifications and performance at the supply chain level. The results indicate that when both, suppliers and buyers, are simultaneously certified the performance levels are significantly higher compared to cases where solely the buyer is certified. However, with regards to ISO 9001 and quality performance our results did not confirm the initial hypothesis one. Being ISO 9001 certified at the supplier and buyer side does not significantly improve the buyers' quality performance. Indeed the results indicate that this has no significant impact on quality performance.

In attempting to explain the support for the two hypotheses related to ISO 14001 (H2) and OSHSAS 18001 (H3) and no support for the hypothesis which investigated ISO 9001 (H1), the institutional theory can be consulted. It could be argued that ISO 9001 certification was viewed by firms as a way to differentiate themselves from competitors. In other words, it was as an order-winning criterion. It has been suggested by Cole (1998) that suppliers may consider quality certification as a primary tool in signalling their credentials to customers. However, given that the standard is well embedded within many organisations, it has become an essential requirement for many customers. Hence, suppliers may be tempted to view the implications of certification from a superficial perspective (Dick 2000), rather than promoting good quality practices. This anomaly between the external certification requirements and the internal organisational quality practices may result in a disconnect, with the standard being implemented simply to comply with institutional pressures.

According to institutional theory, early adopters of a process standard are motivated by the technical efficiency provided by the standard, while late adopters are motivated more by the symbolic value that the standard represents (Meyer and Rowan 1977). As ISO 9001 is increasingly institutionalised, organisations could invoke the socially legitimate goal of ISO 9001 certification without being dedicated to its principles (Westphal et al. 1997). Based on this institutional effect, organizations signal their compliance with institutional demands through symbolic adoption alone and rigorous implementation becomes unnecessary (Guler et al. 2002). The implication is that this would negatively impact performance.

Turning to ISO 14001 and OHSAS 18001, from an institutional perspective, certification helps supplier firms signal legitimacy to major customers (Staw and Epstein 2000). Previous research showed that ISO 9001 was adopted partially for legitimating reasons (Qi et al. 2011); hence it is likely that OHSAS and ISO 14001 certification would have a similar effect. OHSAS 18001 and ISO 14001 are interpreted as a signal of a firm's commitment to health and safety and/or environmental management. Given the increasing demand for organisations to at least appear to meet expectations about health and safety and the environment, such pressure could be a powerful driver toward certification.

However, if this perspective is correct, then it is unlikely that operational performance would remain unchanged as a result of certification to OHSAS 18001 and/or ISO 14001. In other words, if we consider the questionnaire there would be no improvement in environmental performance due to reductions in energy, water use, waste and emissions. Nor would there be health and safety improvements due to reductions in accidents, injuries, ill health and insurance claims. The improvements in performance, when buyers and suppliers have environmental and/or health and safety certification are evident given the positive results for H2 and H3. This study contributes to the overall certification literature by showing that ISO 14001 and OHSAS 18001 that are premised on developing production systems processes that can deliver significant increases in operational performance (Weick et al. 1999).

Another reason that might explain the negative performance results for ISO 9001, is that OHSAS 18001 and ISO 14001 require a much wider stakeholder base relative to ISO 9001. The ISO 9001 standard tends to focus on customers and satisfying their requirements. The other two standards, on the other hand, need to consider the influence of stakeholders from customers to society at large. Given the higher level of scrutiny that this entails, the implication is that this leads to improved performance in terms of environment and occupational health and safety for buyers and suppliers (Castka and Balzarova 2008).

From a management perspective, the results suggest that certified suppliers have an important role to play in the buyer-supplier relationship, particularly with regard to improving environmental and occupational health and safety performance. However, for buyer organisations this may involve active investment in the supply base, particularly smaller suppliers. For example, Nawrocka et al. (2009) reports procurement managers in several organisations providing training and expert knowledge to their supply base. Such measures improved their environmental awareness and prepared the suppliers for more advanced work, such as working towards ISO 14001 or OHSAS 18001 certification. At the same time, companies may wish to focus their resources on more strategically important suppliers during the certification process. Arimura et al. (2011) provides examples of companies investing in modifications to the production line of strategic suppliers into improve environmental and health and safety performance. This is supported by Klassen and Vachon (2003), who found that closer relations to suppliers concerning product related activities were connected to higher tendencies for cooperation on environmental and social issues.

It should also be noted that even though the results would appear to suggest that there is no performance benefit for firms from having ISO 9001 certification, having in place quality management processes and practices should make it easier for buyers and suppliers to implement other standards, such as, ISO 14001 and OSHAS 18001, as they require similar infrastructure and knowledge requirements (Curkovic et al. 2000).

There are a number of limitations with the current study. Firstly, it was country specific and focused on Ireland. Future work should extend the research to consider other jurisdictions. Secondly, it would be useful, given the increased attention to integrated management systems to consider the complementary effect of multiple standards on performance. Thirdly, related to the small sample size it was not feasible to test for the possible confounding implications of industry on our results. However, we do solely include manufacturing firms in our sample. Finally, the study considered three meta-standards, future work could look at other certification programmes, such as ISO 26000 on social responsibility.

## 6 Conclusion

In recent years, firms have implemented quality (ISO 9001), environmental (14001) and occupational health and safety (OHSAS 18001) management standards, in order to remain competitive and meet their stakeholders' objectives. As these standards mature there is a movement by firms to encourage suppliers along the supply chain to consider their implementation. The current study has tried to provide some guidance with regard to the relationship between perceived performance and the three standards outlined above, particularly with regard to its impact on suppliers. The institutional theory perspective suggests that certification may be considered mainly as a signalling device. However, the results only provide support for this contention with regard to ISO 9001 certification, but appear to be less important for ISO 14001 and OHSAS 18001. This implies that these two certifications have had more than just ceremonial benefits and can provide a mechanism for enhanced environmental and health and safety performance.

## References

Abad J, Dalmau I, Vilajosana J (2014) Taxonomic proposal for integration levels of management systems based on empirical evidence and derived corporate benefits. J Clean Prod 78(1):164–173

Abad J, Lafuente E, Vilajosana J (2013) An assessment of the OHSAS 18001 certification process: objective drivers and consequences on safety performance and labour productivity. Saf Sci 60:47–56

Arimura TH, Darnall N, Ganguli R, Katayama H (2016) The effect of ISO 14001 on environmental performance: resolving equivocal findings. J Environ Manage 166:556–566

Arimura TH, Darnall N, Katayama H (2011) Is ISO 14001 a gateway to more advanced voluntary action? The case of green supply chain management. J Environ Econ Manage 61(2):170–182

Arimura TH, Hibiki A, Katayama H (2008) Is a voluntary approach an effective environmental policy instrument? A case for environmental management systems. J Environ Econ Manage 55 (3):281–295

Bansal P, Roth K (2000) Why companies go green: a model of ecological responsiveness. Acad Manage J 43(4):717–736

Barla P (2007) ISO 14001 certification and environmental performance in Quebec's pulp and paper industry. J Environ Econ Manage 53(3):291–306

Benner MJ, Veloso FM (2008) ISO 9000 practices and financial performance: a technology coherence perspective. J Oper Manage 26(5):611–629

Bernardo M, Casadesus M, Karapetrovic S, Heras I (2009) How integrated are environmental, quality and other standardized management systems? An empirical study. J Clean Prod 17 (8):742–750

Blackman A (2012) Does eco-certification boost regulatory compliance in developing countries? ISO 14001 in Mexico. J Regul Econ 42(3):242–263

Boiral O, Henri JF (2012) Modelling the impact of ISO 14001 on environmental performance: a comparative approach. J Environ Manage 99:84–97

Bottani E, Monica L, Vignali G (2009) Safety management systems: performance differences between adopters and non-adopters. Saf Sci 47(2):155–162

Burgen S, Phillips T (2011) Zara Accused in Brazil Sweatshop Inquiry [Online]. Available: http://www.theguardian.com/world/2011/aug/18/zara-brazil-sweatshop-accusation [Accessed 22/04/2016]

Campos LM, de Melo Heizen DA, Verdinelli MA, Miguel PAC (2015) Environmental performance indicators: a study on ISO 14001 certified companies. J Clean Prod 99:286–296

Chang J.I. and Liang C.L. (2009), "Performance evaluation of process safety management systems of paint manufacturing facilities", *Journal of Loss Prevention in the Process Industries,* Vol. 22 No., pp. 398–402

Castka P, Balzarova M (2008) The impact of ISO 9000 and ISO 14000 on standardisation of social responsibility—an inside perspective. Int J Prod Econ 113(1):74–87

Chatzoglou P, Chatzoudes D, Kipraios N (2015) The impact of ISO 9000 certification on firms' financial performance. Int J Oper Prod Manage 35(1):145–174

Chen CC (2005) Incorporating green purchasing into the frame of ISO 14000. J Clean Prod 13 (9):927–933

Chiarini A (2012) Designing an environmental sustainable supply chain through ISO 14001 standard. Manage Environ Qual Int J 24(1):16–33

Choi T, Fan D, Lo CK (2012) A tough pill to swallow? The impact of voluntary occupational health and safety management system on firms' financial performance in fashion and textiles industries. J Fashion Market Manag: Int J 16(2):128–140

Choi TY, Hong Y (2002) Unveiling the structure of supply networks: case studies in Honda, Acura, and DaimlerChrysler. J Oper Manage 20:469–493

Cole RE (1998) Learning from the quality movement: what did and didn't happen and why? Calif Manage Rev 41(1):43–73

Corbett CJ, Montes-Sancho MJ, Kirsch DA (2005) The financial impact of ISO 9000 certification in the United States: AN empirical analysis. Manage Sci 51(7):1046–1059

Curkovic S, Melnyk S, Handfield R, Calantone R (2000) Investigating the linkage between total quality management and environmentally responsible manufacturing. IEEE Trans Eng Manage 47(4):444–464

Curkovic S, Sroufe R (2011) Using ISO 14001 to promote a sustainable supply chain strategy. Bus Strategy Environ 20(2):71–93

Darnall N, Jolley GJ, Handfield R (2008) Environmental management systems and green supply chain management: complements for sustainability? Bus Strategy Environ 17(1):30–45

De Vries HJ, Bayramoglu DK, van der Wiele T (2012) Business and environmental impact of ISO 14001. Int J Qual Reliab Manage 29(4):425–435

Dick G (2000) ISO 9000 certification benefits, reality or myth? The TQM Mag 12(6):365–371

Dick GP, Heras I, Casadesús M (2008) Shedding light on causation between ISO 9001 and improved business performance. Int J Oper Prod Manage 28(7):687–708

ESB (2016) ESB Supplier Charter (online). Available: https://www.esb.ie/docs/default-source/procurement/esb-supplier-charter. Accessed 11 Apr 2016

Fernández-Muñiz B, Montes-Peón JM, Vázquez-Ordás CJ (2009) Relation between occupational safety management and firm performance. Saf Sci 47(7):980–991

Fernández-Muñiz B, Montes-Peón JM, Vázquez-Ordás CJ (2012a) Occupational risk management under the OHSAS 18001 standard: analysis of perceptions and attitudes of certified firms. J Clean Prod 24:36–47

Fernández-Muñiz B, Montes-Peón JM, Vázquez-Ordás CJ (2012b) Safety climate in OHSAS 18001-certified organisations: antecedents and consequences of safety behaviour. Accid Anal Prev 45:745–758

Gray JV, Anand G, Roth AV (2015) The influence of ISO 9000 certification on process compliance. Prod Oper Manage 24(3):369–382

Grimm JH, Hofstetter JS, Sarkis J (2014) Critical factors for sub-supplier management: A sustainable food supply chains perspective. Int J Prod Econ 152(1):159–173

Gualandris J, Kalchschmidt M (2014) Customer pressure and innovativeness: their role in sustainable supply chain management. J Purch Supply Manage 20(2):92–103

Gualandris J, Kalchschmidt M (2016) Developing environmental and social performance: the role of suppliers' sustainability and buyer–supplier trust. Int J Prod Res 54(8):1–17

Guler I, Guillen MF, Macpherson JM (2002) Global competition, institutions, and the diffusion of organizational practices: the international spread of the ISO 9000 quality certificates. Adm Sci Q 47(2):207–232

Han SB, Chen SK, Ebrahimpour M (2007) The impact of ISO 9000 on TQM and business performance. J Bus Econ Stud 13(2):1–23

Handfield R, Sroufe R, Walton S (2005) Integrating environmental management and supply chain strategies. Bus Strategy Environ 14(1):1–19

Hohnen P, Hasle P (2011) Making work environment auditable—a 'critical case' study of certified occupational health and safety management systems in Denmark. Saf Sci 49:1022–1029
Hora M, Bapuji H, Roth AV (2011) Safety hazard and time to recall: The role of recall strategy, product defect type, and supply chain player in the US toy industry. J Oper Manage 29(7–8):766–777
Huo B, Han Z, Prajogo D (2014) The effect of ISO 9000 implementation on flow management". Int J Prod Res 52:6467–6481
Hutcheson G, Sofroniou N (1999) The multivariate social scientist: introductory statistics using generalized linear models. Sage Publications, Thousand Oaks
Hwang I, Radhakrishnan S, Su L (2006) Vendor certification and appraisal: implications for supplier quality. Manag Sci 52(10):1472–1482
ISO Central Secretariat (2009a) Selection and use of the ISO 9000 family of standards. http://www.iso.org/iso/iso_9000_selection_and_use-2009.pdf. 22 Sept 2015
ISO Central Secretariat (2009b) Environmental management – The ISO 14000 family of international standards. http://www.iso.org/iso/theiso14000family_2009.pdf. 22 Sept 2015
King AA, Lenox MJ, Terlaak A (2005) The strategic use of decentralized institutions: Exploring certification with the ISO 14001 management standard. Acad Manag J 48(6):1091–1106
Klassen R, Vachon S (2003) Collaboration and evaluation in the supply chain: the impact on plant-level environmental investment. Prod Oper Manage 12(3):336–352
Kline P (1979) Psychometrics and psychology. Acaderric Press, London
Koplin J, Seuring S, Mesterharm M (2007) Incorporating sustainability into supply management in the automotive industry—the case of the Volkswagen AG. J Clean Prod 15(11–12):1053–1062
Krause DR, Handfield RB, Tyler BB (2007) The relationships between supplier development, commitment, social capital accumulation and performance improvement. J Oper Manage 25:528–545
Lafuente E, Bayo-Moriones A, García-Cestona M (2010) ISO-9000 certification and ownership structure: effects upon firm performance. Br J Manage 21(3):649–665
Lam PT, Chan EH, Chau C, Poon C, Chun K (2011) Environmental management system vs green specifications: how do they complement each other in the construction industry? J Environ Manage 92:788–795
Lambert DM, Cooper MC (2000) Issues in supply chain management. Ind Mark Manage 29:65–83
Lee AH, Kang HY, Hsu CF, Hung HC (2009) A green supplier selection model for high-tech industry. Expert Syst Appl 36(4):7917–7927
Levine DI, Toffel MW (2010) Quality management and job quality: how the ISO 9001 standard for quality management systems affects employees and employers. Manage Sci 56(6):978–996
Lo CK, Pagell M, Fan D, Wiengarten F, Yeung AC (2014) OHSAS 18001 certification and operating performance: the role of complexity and coupling. J Oper Manage 32:268–280
Lo CK, Wiengarten F, Humphreys P, Yeung AC, Cheng T (2013) The impact of contextual factors on the efficacy of ISO 9000 adoption. J Oper Manage 31:229–235
Lund-Thomsen P, Lindgreen A (2014) Corporate social responsibility in global value chains: where are we now and where are we going? J Bus Ethics 123(1):11–22
Maloni MJ, Brown ME (2006) Corporate social responsibility in the supply chain: an application in the food industry. J Bus Ethics 68(1):35–52
Marshall D, McCarthy L, McGrath P, Claudy M (2016) Going above and beyond: How sustainability culture and entrepreneurial orientation drive social sustainbility supply chain practice adoption. Supply Chain Manage Int J 20(4):434–454
Melnyk SA, Sroufe RP, Calantone RJ (2003) A model of site-specific antecedents of ISO 14001 certification. Prod Oper Manage 12(3):369–385
Meyer JW, Rowan B (1977) Institutionalized organizations: formal structure as myth and ceremony. Am J Sociol 83(2):340–363
Montabon F, Sroufe R, Narasimhan R (2007) An examination of corporate reporting, environmental management practices and firm performance. J Oper Manage 25(5):998–1014
Morris PW (2006) ISO 9000 and financial performance in the electronics industry. J Am Acad Bus 8(2):227–235

Nawrocka D, Brorson T, Lindhqvist T (2009) ISO 14001 in environmental supply chain practices. J Clean Prod 17(16):1435–1443

Nishitani K, Kaneko S, Fujii H, Komatsu S (2012) Are firms' voluntary environmental management activities beneficial for the environment and business? An empirical study focusing on Japanese manufacturing firms. J Environ Manage 105:121–130

Norrman A, Jansson U (2004) Ericsson's proactive supply chain risk management approach after a serious sub-supplier accident. Int J Phys Distrib Logist Manage 34(5):434–456

Nunnally JC (1978) Psychometric theory. McGraw Hill, New York

OHSAS Project Group (2011) Results of the survey into the availability of OH&S standards and certificates, up until 2009-12-31. http://ohsas18001expert.com/wp-content/uploads/2011/05/2009-OHSAS-Certificates-Survey-Results.pdf. Assessed on 07 Feb 13

Pagell M, Shevchenko A (2014) Why research in sustainable supply chain management should have no future. J supp chain manag 50(1):44–55

Pagell M, Wu Z, Wasserman ME (2010) Thinking differently about purchasing portfolios: an assessment of sustainable sourcing. J Supply Chain Manage 46(1):57–73

Podsakoff PM, MacKenzie SB, Lee J, Podsakoff NP (2003) Common method bias in behavioral research: a critical review of the literature and recommended remedies. J Appl Psychol 88 (5):879–903

Potoski M, Prakash A (2005) Covenants with weak swords: ISO 14001 and facilities' environmental performance. J Policy Anal Manage 24(4):745–769

Prajogo D, Huo B, Han Z (2012) The effects of different aspects of ISO 9000 implementation on key supply chain management practices and operational performance. Supply Chain Manage Int J 17(3):306–322

Psomas EL, Pantouvakis A, Kafetzopoulos DP (2013) The impact of ISO 9001 effectiveness on the performance of service companies. Manage Serv Qual Int J 23(2):149–164

Qi G, Zeng S, Yin H, Lin H (2013) ISO and OHSAS certifications: how stakeholders affect corporate decisions on sustainability. Manage Decis 51:1983–2005

Qi Y, Zhao X, Sheu C (2011) The impact of competitive strategy and supply chain strategy on business performance: the role of environmental uncertainty. Decision Sci 42(2):371–389

Rosenzweig ED, Roth AV (2004) Towards a theory of competitive progression: evidence from high-tech manufacturing. Prod Oper Manag 13(4):354–368

Sharma DS (2005) The association between ISO 9000 certification and financial performance. Int J Account 40(2):151–172

Shin H, Collier DA, Wilson DD (2000) Supply management orientation and supplier/buyer performance. J oper manag 18(3):317–333

Singh PJ, Power D, Chuong SC (2011) A resource dependence theory perspective of ISO 9000 in managing organizational environment. J Oper Manage 29:49–64

Srivastav AK (2011) ISO 9000 as an organisation development intervention. TQM J 23(3):313–325

Staw BM, Epstein LD (2000) What band wagons bring: effects of popular management techniques on corporate performance, reputation, and CEO pay. Adm Sci Q 45(3):523–556

Starke F, Eunni RV, Manoel Martins Dias Fouto N, Felisoni de Angelo C (2012) Impact of ISO 9000 certification on firm performance: evidence from Brazil. Manage Res Rev 35(10):974–997

Testa F, Rizzi F, Daddi T, Gusmerotti NM, Frey M, Iraldo F (2014) EMAS and ISO 14001: the differences in effectively improving environmental performance. J Clean Prod 68:165–173

Vachon S, Klassen RD (2006) Green project partnership in the supply chain: the case of the package printing industry. J Clean Prod 14:661–671

Vanpoucke E, Vereecke A, Boyer KK (2014) Triggers and patterns of integration initiatives in successful buyer–supplier relationships. J Oper Manage 32:15–33

Vinodkumar M, Bhasi M (2011) A study on the impact of management system certification on safety management. Saf Sci 49(3):498–507

Weick K, Sutcliffe K, Obstfeld D (1999) Organizing for high reliability: processes of collective mindfulness. Res Organ Behav 21:23–81

Westphal JD, Gulati R, Shortell SM (1997) Customization or conformity? An institutional and network perspective on the content and consequences of TQM adoption. Adm Sci Q 42 (2):366–394

Wiengarten F, Pagell M, Fynes B (2013) ISO 14000 certification and investments in environmental supply chain management practices: identifying differences in motivation and adoption levels between Western European and North American companies. J Clean Prod 56:18–28

Wilhelm MM, Blome C, Bhakoo V, Paulraj A (2016) Sustainability in multi-tier supply chains: Understanding the double agency role of the first-tier supplier. J Oper Manage 41:42–60

Zobel T (2013) ISO 14001 certification in manufacturing firms: a tool for those in need or an indication of greenness? J Clean Prod 43:37–44

Zutshi A, Sohal AS (2004) Adoption and maintenance of environmental management systems: critical success factors. Manage Environ Qual Int J 15(4):399–419

CPSIA information can be obtained
at www.ICGtesting.com
Printed in the USA
LVOW10*0618231017
553423LV00008B/73/P

9 783319 656748